WiMAX

Taking Wireless to the MAX

WiMAX

Taking Wireless to the MAX

Deepak Pareek

Auerbach Publications
Taylor & Francis Group
Boca Raton   New York

Auerbach Publications is an imprint of the
Taylor & Francis Group, an informa business

Published in 2006 by
Auerbach Publications
Taylor & Francis Group
6000 Broken Sound Parkway NW, Suite 300
Boca Raton, FL 33487-2742

© 2006 by Taylor & Francis Group, LLC
Auerbach is an imprint of Taylor & Francis Group

International Standard Book Number-10: 0-8493-7186-4 (Hardcover)
International Standard Book Number-13: 978-0-8493-7186-8 (Hardcover)

Taylor & Francis Group
is the Academic Division of Informa plc.

Visit the Taylor & Francis Web site at
http://www.taylorandfrancis.com

and the Auerbach Publications Web site at
http://www.auerbach-publications.com

Contents

Preface

Knowledge workers today need to be increasingly agile to adapt to the dynamic competitive environment and the changing needs of their professional and social lifestyle. Information has become a vital tool as connectivity is a necessity now. An increasing demand for access to information anytime, anywhere, has led to an explosive growth of both access and wireless technologies.

These modern technologies are revolutionizing the way we work, play, and interact. It won't be an exaggeration if we suggest these disruptive technologies are altering the way we live, making life better for all of us. More interesting to note is that with every passing day these disruptions are becoming more frequent. This trend has created new competitive threats as well as new opportunities in every walk of life.

The telecommunications industry is finding itself most affected by what is happening. The underlying human behavioral trait responsible for this unprecedented but welcomed activity is the need for communication. Because the telecommunications industry is now more mature then ever, players in this sphere are on their toes trying to find new solutions and technologies, new ways to add value to their existing products and services, to gain a competitive advantage, and to increase customer loyalty while also attracting new, high-value clients.

Think of the possibilities that these access and wireless technologies offer to a wide range of customers. From having the freedom to pay bills while stuck in traffic, to receiving notification of a change in professional tax rates while having lunch, and to taking French lessons while returning home on the bus or train, the convenience, productivity, and time-saving benefits of these advancements are huge.

Further, these advancements can save the lives of millions of people living in underdeveloped parts of the globe by providing remote health care services and emergency or distress information regarding possible typhoons, floods, or tsunamis. Even an underserved poor child living in the sub-Sahara can read details of the latest experiments in space science or biotechnology conducted in California or at Oxford. The potential of these technologies is phenomenal.

The challenge, then, is how to turn these possibilities into realities, to provide solutions that enable anywhere, anytime access to information and applications at low cost and with a small investment.

WiMAX: Taking Wireless to the MAX is a step in this direction as it demystifies WiMAX (Worldwide Interoperability for Microwave Access), a technology for broadband wireless access (BWA) that challenges the T1, DSL, or cable modems generally used today, with their physical cables called "landlines."

WiMAX, an evolving standard for point-to-multi-point wireless networking, works for the "last mile" in the same way that WiFi "hot spots" work for the last 100 feet of networking within a building or a home. However, in the case of WiMAX, the last mile literally means *miles* as, theoretically, it can serve up to 30 to 31 miles.

This book examines the technologies, trends, evolution, application, models, cases and standards for wireless broadband with a focus on WiMAX. The outcome of dedicated research and analysis efforts by the author and his team, with support from many distinguished authorities in this field, provides strategic insights and also makes recommendations to different actors related to WiMAX.

The first chapter is an introduction and examines various phases of the evolution of the wireless, broadband, and wireless broadband landscape. It also covers various technologies and topologies in these areas. The chapter gives an overview of things to expect in later parts of the book. Further it sets the tone for the journey toward untouched wireless heights — the WiMAX.

The second chapter "Setting the Stage" describes the direction for the later part of the book. It discusses several issues and challenges leading to the need for WiMAX. One of the key areas covered is the digital divide and knowledge society.

The third chapter "Telecommunication: A Connecting Mechanism" provides insight into telecommunication landscape. It takes a close look at various transmission and access technologies, and it examines various aspects of data versus voice network debates.

The fourth chapter "The Internet Takes Off" discusses the birth and growth of the Internet, and its impact on society and business. It also provides detailed commentary on underlying technologies, on technology areas (backhaul, distribution, and access), and on technology advancements that were responsible for the evolution and proliferation of the Internet.

The fifth chapter "The Broader the Better" is a comprehensive resource covering broadband evolution. It takes a close look at the factors and applications behind the surge of broadband and the impact of broadband across multiple segments, and it identifies compelling reasons behind the hype created by broadband. The chapter discusses various technology options available for delivering broadband and presents comparative analysis of various broadband technologies.

The sixth chapter "Wired versus Wireless" gives an assessment of wireless and wireline technologies. It highlights the forces behind the wireless explosion and directs attention toward future evolution.

The seventh chapter "Broadband Unwired" presents one of the most dynamic, promising, and exciting technology areas. It provides complete coverage of BWA technology from fixed wireless to Beyond 3G, and WLAN, WMAN, and WPAN. This chapter also introduces the concept of personal broadband.

The eighth chapter "Understanding the Technology" provides a detailed discussion of WiMAX architecture and deployment, modulation techniques, and network topology. This chapter also examines key wireless standards and touches on spectrum issues relevant to the scope of the book.

The ninth chapter "Surveying the Landscape" explores cost economics of WiMAX, market segments and scope, technical specifications, and features that make WiMAX a revolutionary development. The chapter also gives a brief account of WiMAX applications and the value it adds to various stake holders. It also discusses in detail WiMAX as an enabler of convergence, while it takes a close look at the impact of Moore's law on communication landscape.

The tenth chapter "Identifying the Market" discusses WiMAX's place in the wireless family, the WiMAX value chain, the present status of WiMAX development, and future opportunities. It provides a detailed overview of the possible challenges WiMAX will face in future, including RF interference, spectrum issues, infrastructure placement, and regulatory environment. Also included in this chapter are accounts of evolution, functions, operations, and the role of the WiMAX Forum, as well as an in-depth insight into the process followed by the WiMAX Forum for profile creation, test specification creation, conformance testing, interoperability testing, and the certification process.

The eleventh chapter "Predicting the Future" is an analysis that enhances the understanding about the future of broadband wireless with WiMAX as the focal point, while providing global forecasts and trends.

The twelfth chapter "Analyzing the Model" gives an understanding about various possible business models and provides an innovative hypothesis for WiMAX. A detailed pragmatic SWOT analysis for each model is another highlight of this chapter.

The thirteenth chapter "Planning the Strategy" reveals the key to success in the dynamic BWA market. It briefly discusses options available to all stake holders in WiMAX initiative, suggests methodology for deciding future plans of action, and provides answers to the million dollar question *To WiMAX or Not to WiMAX?* Further, it guides players on collective efforts as well as on individual efforts needed to make WIMAX technology deployment a reality for practical applications in everyday life.

The book's concluding chapter collates the essence of knowledge and learning presented. The book also incorporates some detailed readings on different topics that have been touched upon in the main text but were not covered in the interest of information required by a more general audience.

WiMAX — Just wait and watch.

About the Author

Focusing on creative, forward-looking strategies, Deepak Pareek is an author, coach, and speaker. His formal education was obtained in the fields of engineering (telecom and IT), business (strategic management), and economics (international trade).

An expert in business planning, strategy, and analysis, he has a decade of hands-on experience in multiple wireless and telecommunication technologies. He has extensive global management and technology consulting experience, and has worked at the top of the summit organizations, assisting them by providing futuristic vision and technology road maps.

He speaks regularly on topics related to management, policy, leadership, and technology at globally reputed platforms. He is also the author of many papers, concept notes, analytics, columns, articles, and reports.

His other published books are *Vision of Indian Telecom Sector 2020*, referred to by the World Trade Organization and ITU as a case for tariff negotiation for Southeast Asia, and the *Handbook of Telecommunication Management*.

He describes himself as approaching challenges with a "never say die" attitude and a desire to "contribute in improving the world around us."

Chapter 1

Introduction

The Internet will break down national borders and lead to world peace ... children are not going to know what nationalism is.

Nicholas Negroponte,
Massachusetts Institute of Technology, 1997

The communications landscape is changing dramatically under the increasing pressure of rapid technological development and intensifying competition. The most significant development in the communications industry in the past ten years has been the dramatic increase in network capabilities and the subsequent fall in communications pricing.

The effects of this revolution have been felt in almost every sector including banking, investment, healthcare, real estate, education, trading, manufacturing, governance, and law. The empowering capabilities of advanced communication technologies will certainly be the pivotal force shaping economies and societies over the next few years.

In less than one generation, the distinctions among telephone, broadcast, cable, satellite, wireless, and information services have all but disappeared as broadband, and wireless Internet technologies subsume all modes of news, entertainment, data, and voice transmissions.

The vision of ubiquitous access to information, anytime, anywhere, is becoming a reality, enabled by rapidly emerging wireless communications technologies with coverage that ranges from a few inches to many miles. These technologies have the potential to dramatically change society. The age of untethered computing is here. Productivity is no longer limited to areas with network connections. Users can now move from place to place, computing when and where they want.

Give Me More: The Need for Broadband

Although the 1970s and 1980s will be remembered as the information age and the 1990s will undoubtedly be singled out in history as the beginning of the Internet age, the first decades of the twenty-first century may become the broadband age, or even better, the age of convergence. The advent of the networked computer was truly revolutionary in terms of information processing, data sharing, and data storage. In the 1990s, the Internet's influence was even more revolutionary in terms of communications and furthering the progress of data sharing from the personal level to the global enterprise level.

Today, broadband elements such as fiber optics, wireless access, and cable modems provide very-high-speed access to information and media of all types via corporate networks and the Internet, creating an "always-on" environment. The result will eventually be a widespread convergence of entertainment, telephony, and computerized information: data, voice and video, delivered to a rapidly evolving array of Internet appliances, personal digital assistants (PDAs), wireless devices (including cellular telephones), and desktop computers.

The broadband market continues to be a dynamic sector as the competitive landscape and consumer demand for new communication services continue to evolve. Driven by the need to find new sources of revenue, service providers are looking for ways to unleash the potential of broadband networks. We will cover these aspects in more detail in later chapters.

Going Wireless: Enabling the Revolution

There is no doubt that the world is going wireless faster and more widely than anyone might have expected. The present-day reality is that billions of people will gain high-speed wireless Internet access within the next decade. The transition to wireless really began during the Internet revolution. What started as an exchange mechanism for electronic data has sparked worldwide demand for anytime, anywhere computing and communications.

The impetus for the broadband wireless revolution is coming from consumers and businesses worldwide who increasingly expect to enjoy wireless computing and communications anytime, anywhere. It will require a plethora of solutions, technologies, components, platforms, infrastructure, and services to meet this demand. Not since the early days of the Internet era have there been so many new revenue-generating opportunities.

It should be noted that even though more and more people are getting connected without wires, this does not mean that wired access will disappear. In fact, wired technologies will continue to be important, as it is difficult to imagine the entire world's computing infrastructure operating without Gigabit Ethernet. Ethernet and other wired technologies such as InfiniBand and Fibre Channel play a vital behind-the-scenes role in the infrastructure enabling wireless connectivity as well as providing the fastest available connection option to the mobile platform users.

Broadband Goes Wireless

Broadband wireless is a continuum of coexisting, overlapping technologies that enable wireless high-speed communications. Wireless Fidelity (Wi-Fi), Worldwide Interoperability for Microwave Access (WiMAX), third-generation (3G), and Ultrawideband (UWB) technologies each are necessary for the global wireless infrastructure required to deliver high-speed communications and Internet access worldwide. Whereas Wi-Fi is ideal for isolated islands of connectivity, WiMAX and 3G are needed for long-distance wireless "canopies."

Meanwhile, WiMAX and 3G are both required because their optimum platforms differ: WiMAX works best for computing platforms, such as laptops, whereas 3G is best for mobile devices such as PDAs and cell phones. UWB offers very-short-range connectivity, perfect for the home entertainment environment or wireless USB. In short, each technology is important for different reasons.

All of the wireless networks will get built for different usages, with some overlap at the edges. But most importantly, the technologies will coexist, creating more robust solutions that will enable a host of new and exciting possibilities. In essence, the term broadband wireless encompasses the full range of wireless technologies and applications — both fixed and mobile.

Understanding Wireless Networks

A wireless network is a radio access system designed to provide location-independent network access between computing devices. It is usually implemented as the final link between an existing network and a group of client computers, giving these users wireless access to the full resources and services of the network across a specific distance, depending upon technology used.

Wireless networking technologies range from global voice and data networks (which allow users to establish wireless connections across long distances) to infrared light and radio frequency technologies that are optimized for short-range wireless connections. Devices commonly used for wireless networking include portable computers, desktop computers, handheld computers, PDAs, cellular phones, pen-based computers, and pagers. Wireless technologies serve many practical purposes. For example, mobile users can use their cellular phone to access e-mail. Travelers with portable computers can connect to the Internet through base stations installed in airports, railway stations, and other public locations. At home, users can connect devices on their desktop to synchronize data and transfer files.

Defining Standards

Before broadband wireless can deliver on its promises, the entire communications industry must embrace the notion that coexisting, standards-based technologies are the right strategy. In addition, those standards must be delivered via modular, cost-effective platforms that will enable greater innovation and interoperability. As the industry works together to conform to standards, they also create value for end users. Some key advantages are the following:

- Common design criteria will allow products from multiple vendors to work together in a solution.
- Broader market enables mass production, leading to lower costs and worldwide economies of scale.
- Proliferation of mobile computing devices built on common architectures creates quick and easy opportunities to launch new services, leading to faster time to profit and quicker time to market.

- Faster pace of innovation when multiple vendors compete for revenue opportunities.
- Greater emphasis on service capabilities and applications as vendors focus on differentiation; reduced reliance on proprietary components and designs.
- Standards compliance and interoperability will create new worldwide market segments for platforms and solutions.

To lower costs, ensure interoperability, and promote the widespread adoption of wireless technologies, organizations such as the Institute of Electrical and Electronics Engineers (IEEE), Internet Engineering Task Force (IETF), Wireless Ethernet Compatibility Alliance (WECA), and the International Telecommunication Union (ITU) are participating in several major standardization efforts. For example, IEEE working groups are defining how information should be transferred from one device to another (whether radio waves or infrared light should be used, for example) and how and when a transmission medium should be used for communications. In developing wireless networking standards, organizations such as the IEEE address power management, bandwidth, security, and issues that are unique to wireless networking. Other organizations such as Wi-Fi Alliance, WiMAX Forum, and many more are continuously working toward breaking new ground in communications, especially in wireless domain.

Wireless Network Types

As with wired networks, wireless networks can be classified into different types based on the distances over which data can be transmitted. Wireless networks are becoming more pervasive, accelerated by new wireless communications technologies, inexpensive wireless equipment, and broader Internet access availability. These networks are transforming the way people use computers and other personal electronics devices at work, home, and when traveling. There are many wireless communications technologies that can be differentiated by frequency, bandwidth, range, and applications. In this section, we survey these technologies, which can be broadly organized into the four categories depicted in Figure 1.1. These categories range from wireless wide area networks (WWANs), which cover the widest geographic area, to wireless personal area networks (WPANs), which cover less than 10 m.

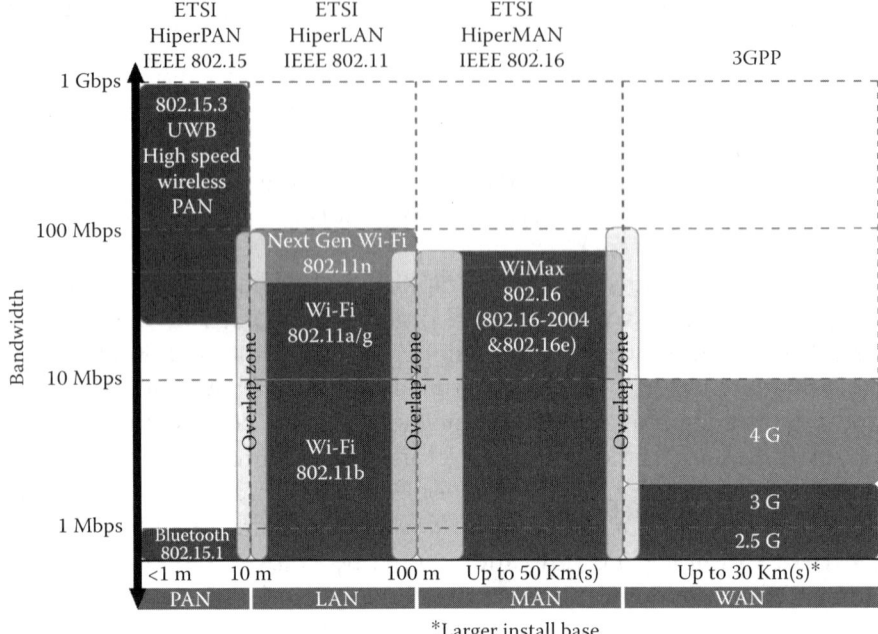

Figure 1.1 Wireless access technologies.

Wireless Wide Area Networks

WWAN technologies enable users to establish wireless connections over remote public or private networks. These connections can be maintained over large geographic areas, such as cities or countries, through the use of multiple antenna sites or satellite systems maintained by wireless service providers. Current WWAN technologies are known as second-generation (2G) systems. Key 2G systems include Global System for Mobile Communications (GSM), Cellular Digital Packet Data (CDPD), and Code Division Multiple Access (CDMA). Efforts are under way to transition from 2G networks, some of which have limited roaming capabilities and are incompatible with each other, to 3G technologies that would follow a global standard and provide worldwide roaming capabilities. The ITU is actively promoting the development of a global standard for 3G.

Wireless Metropolitan Area Networks

Wireless metropolitan area network (WMAN) technologies enable users to establish wireless connections between multiple locations within a metropolitan area (for example, between multiple office buildings in a city or on a university campus), without the high cost of laying fiber or copper cabling and leasing lines. In addition, WMANs can serve as backups for wired networks should the primary leased lines for wired networks become unavailable. WMANs use either radio waves or infrared light to transmit data. Broadband wireless access networks, which provide users with high-speed access to the Internet, are in increasing demand. Although different technologies, such as the Multi-Channel Multi-Point Distribution Service (MMDS) and the Local Multi-Point Distribution Services (LMDS), are being used, the IEEE 802.16 working group for broadband wireless access standards is still developing specifications to standardize development of these technologies.

Wireless Local Area Networks

Wireless local area network (WLAN) technologies enable users to establish wireless connections within a local area (for example, within a corporate or campus building, or in a public space, such as an airport). WLANs can be used in temporary offices or other spaces where the installation of extensive cabling would be prohibitively expensive, or to supplement an existing LAN so that users can work at different locations within a building at different times. WLANs can operate in two different ways. In infrastructure WLANs, wireless stations (devices with radio network cards or external modems) connect to wireless access points that function as bridges between the stations and the existing network backbone. In peer-to-peer (ad hoc) WLANS, several users within a limited area, such as a conference room, can form a temporary network without using access points if they do not require access to network resources.

In 1997, IEEE approved the 802.11 standard for WLANs, which specifies a data transfer rate of 1 to 2 Mbps. Under 802.11b, which is emerging as the new dominant standard, data is transferred at a maximum rate of 11 Mbps over a 2.4 GHz frequency band. Another newer standard is 802.11a, which specifies data transfer at a maximum rate of 54 Mbps over a 5 GHz frequency band.

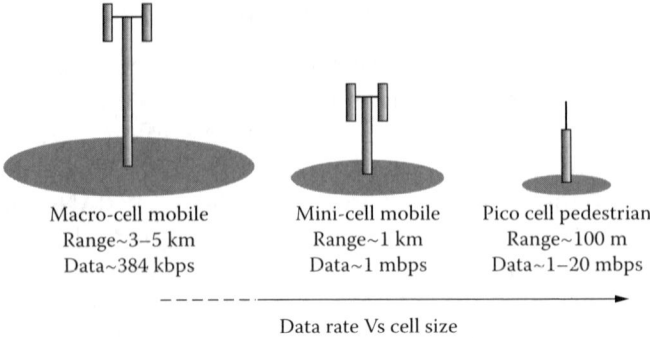

Macro-cell mobile
Range~3–5 km
Data~384 kbps

Mini-cell mobile
Range~1 km
Data~1 mbps

Pico cell pedestrian
Range~100 m
Data~1–20 mbps

Data rate Vs cell size

Figure 1.2   Cell size versus data rate.

Wireless Personal Area Networks

WPAN technologies enable users to establish ad hoc, wireless communications for devices (such as PDAs, cellular phones, or laptops) that are used within a personal operating space (POS). A POS is the space surrounding a person, up to a distance of 10 m. Currently, the two key WPAN technologies are Bluetooth and infrared light. Bluetooth is a cable replacement technology that uses radio waves to transmit data to a distance of up to 30 ft. Bluetooth data can be transferred through walls, pockets, and briefcases. Technology development for Bluetooth is driven by the Bluetooth Special Interest Group (SIG), which published the Bluetooth version 1.0 specification in 1999. Alternatively, to connect devices at a very close range (1 m or less), users can create infrared links.

To standardize the development of WPAN technologies, IEEE has established the 802.15 working group for WPANs. This working group is developing a WPAN standard, based on the Bluetooth version 1.0 specification. Key goals for this draft standard are low complexity, low power consumption, interoperability, and coexistence with 802.11 networks.

Key Wireless Technologies

UWB

WPANs are very small networks within a confined space, such as an office workspace or a room within a home. UWB technologies, offering WPAN users a much faster, short-distance connection, are currently under development.

Wi-Fi

WLANs have broader range than WPANs, typically confined within office buildings, restaurants, stores, homes, etc. WLANs are gaining in popularity, fueled in part by the availability of devices optimized for wireless computing.

Wi-Fi encompasses a family of specifications within the IEEE 802.11 standard. These include 802.11b (the most popular, at 11 Mbps, with a typical range of up to 300 ft), 802.11a (54 Mbps, but with a shorter range than 802.11b) and 802.11g (combining the speed of "a" with the range of "b").

WiMAX

WMANs cover a much greater distance than WLANs, connecting buildings to one another over a broader geographic area. The emerging WiMAX technology (802.16d today and 802.16e in the near future) will further enable mobility and reduce reliance on wired connections.

WiMAX is the new shorthand term for IEEE Standard 802.16, also known as *Air Interface for Fixed Broadband Wireless Access Systems*. The initial version of the 802.16 standard, approved by the New-York-based IEEE in 2002, operates in the 10-to-66-GHz frequency band and requires line-of-sight towers.

The 802.16a extension, ratified in March 2003, does not require line-of-sight transmission and allows use of lower frequencies (2 to 11 GHz), many of which are unregulated. It boasts a 31 mi range and 70 Mbps data transfer rates that can support thousands of users. Additional 802.16 standards are in the works and will cover:

- 802.16b — Quality of service
- 802.16c — Interoperability, with protocols and test-suite structures
- 802.16d — Fixing things not covered by 802.11c, which is the standard for developing access points
- 802.16e — Support for mobile as well as fixed broadband

Cellular Technologies and the Emergence of 3G

WWANs are the broadest-range wireless networks and are most widely deployed today in the cellular voice infrastructure, although they also have the ability to transmit data. Wireless provider networks and WANs that have been considered cellular voice are changing to carry data.

There are three predominant digital cellular wireless technologies employed globally: Time Division Multiple Access (TDMA), CDMA, and GSM.

TDMA is the oldest, simplest digital cellular technology. It is analogous to time division multiplexing in wired networks. TDMA does not have a technological future. Major wireless and eventually all services are in the process of replacing it with other technologies. Next-generation cellular services based on various 3G technologies will significantly improve WWAN communications.

The official 3G wireless standard, known as IMT-2000, has two primary incompatible variations: CDMA2000 and WCDMA, also known as Universal Mobile Telephone Service (UMTS).

GSM is a digital mobile telephone system that is widely used in Europe and other parts of the world. GSM uses a variation of TDMA and is the most widely used of the three digital wireless telephone technologies (TDMA, GSM, and CDMA).

CDMA's progression to 3G technology is easier than GSM's because the underlying technologies are the same. CDMA2000 technology can be deployed on the same systems and radio transmitters as legacy CDMA. This eases the network transition (because both networks operate simultaneously) and handset transition (both legacy 2G handsets and 3G handsets will operate with the same network).

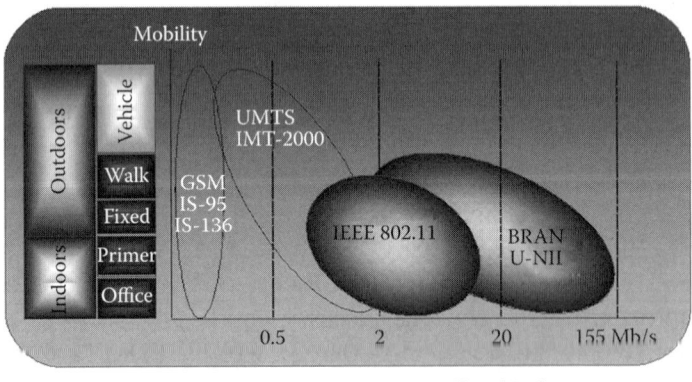

GSM = Global System for
Mobile communications

UMTS = Universal Mobile
Telecommunications System

Bran = Broadband Radio
Access Networks

U-NII = Unlicensed-National
Information Infrastructure

IMT-2000 = International Mobile
Telecommunications-2000

Figure 1.3   Technologies and mobility.

Dynamics of Wireless Technologies

Since the turn of the millennium, wireless networks have proliferated. The popularity of wireless networking has grown very quickly because of effective standardization. Wi-Fi has freed us, enabling us to move around our offices and public places with our laptops and handhelds while retaining instant, unencumbered access to our companies' intranets and the Internet. WiMAX is the next step on the road to a wireless world, extending broadband wireless access to new locations and over longer distances, as well as significantly reducing the cost of bringing broadband to new areas.

Among the promises of WiMAX is that it could offer the solution to what is sometimes called the *last mile* problem, referring to the expense and time needed to connect individual homes and offices to trunk lines for communications. WiMAX promises a wireless access range of up to 31 mi, compared with Wi-Fi's 300 ft and Bluetooth's 30 ft. WiMAX has been designed from the beginning to be compatible with European standards — something that did not happen with 802.11a and delayed its adoption.

The overall concept of metropolitan area wireless networking, as envisioned in 802.16, begins with what is called *fixed wireless*. Here, a backbone of base stations is connected to a public network, and each station supports hundreds of fixed subscriber stations, which can be both public Wi-Fi hot spots and firewalled enterprise networks. The base stations would use the media access control layer defined in the standard, a common interface that makes the networks interoperable, and would allocate uplink and downlink bandwidth to subscribers according to their needs, on an essentially real-time basis.

Later in the development cycle, with 802.16e, WiMAX is expected to support mobile wireless technology — that is, wireless transmissions directly to mobile end users. This will be similar in function to the General Packet Radio Service and the one times radio transmission technology (1×RTT) offered by phone companies.

Following on the heels of WiMAX is another standard, IEEE 802.20, which addresses wide area wireless networks and is currently under development; no products supporting 802.20 are expected before 2006.

As computing and communications converge on broadband wireless platforms and technologies, demand for true mobility will soar. When that happens, industry leaders must be ready to deliver the technologies, infrastructure, devices, and services that enable users to stay connected through the best available technology even as they move about — across the room, across the street, and across the globe. This is the

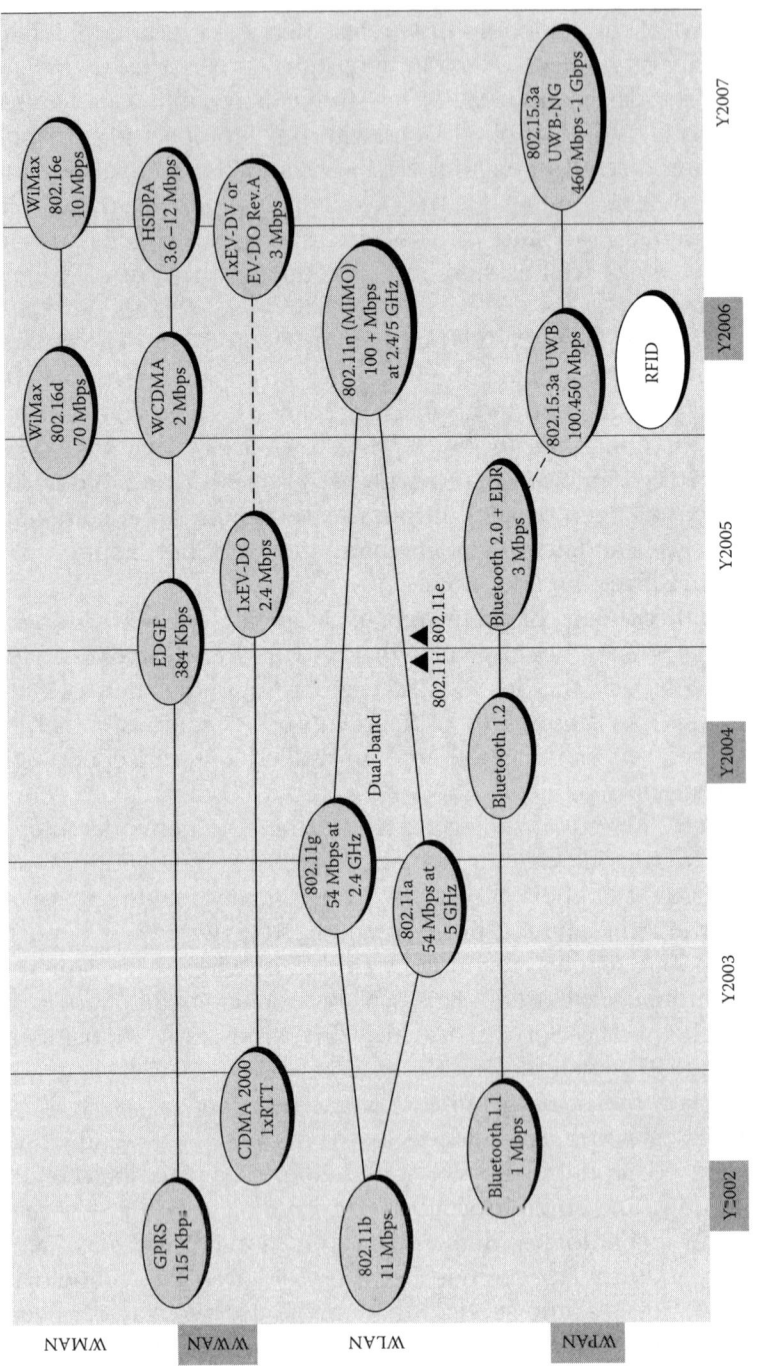

Figure 1.4 Evolution of wireless access technology.

always-best-connected goal, toward which broadband technologies such as 3G, UWB, Wi-Fi, and WiMAX will work synergistically to deliver secure data with anytime, anywhere connectivity. These overlapping wireless networks will offer users choices for the best possible connection. In fact, the mobility enabled by wireless technology necessitates overlap between networks and coexistence among technologies — wired and wireless.

All high-speed wireless technologies (3G, Wi-Fi, WiMAX, and UWB) will coexist, working in tandem to meet service provider and customer needs for truly mobile computing and communications across the globe. No single technology will become dominant or ubiquitous — they all meet unique user requirements in a wirelessly connected world. In fact, the most robust wireless solutions will use a combination of technologies to enable increased mobility and eventually seamless roaming.

It is this demand for mobility that will continue to fuel convergence and transform the communications industry. To that end, industry leaders are developing new wireless standards that will expand and extend the reach of wireless networks across the globe. Meanwhile, carriers have slowed expansion of the fiber network in anticipation of new wireless technologies. And engineers are focusing new development on the products and services that will enable broadband wireless communications on a wide scale.

Opportunity for Wireless Technologies

Broadband wireless presents the most viable opportunity to improve communications for the 1 billion people who currently enjoy Internet access, and to connect the projected 5 billion new users. So much momentum is being generated around wireless communications that the next decade has been designated by some of the top industry leaders as the Broadband Wireless Era.

There are several options for wireless communication currently available through service providers. The majority of users connect via cellular connections, either using the GSM family of networks (GSM, GPRS, EDGE, UMTS), or the CDMA family of cellular networks (CDMA, CDMA 2000, 1 × RTT, EV-DO, EV-DV).

Additionally, various WPAN technologies are emerging as well. Bluetooth is well on its way to become the most widely deployed WPAN technology. Among handsets and other devices, devices that are Bluetooth-enabled would number nearly 300 million by 2007. Looking a few

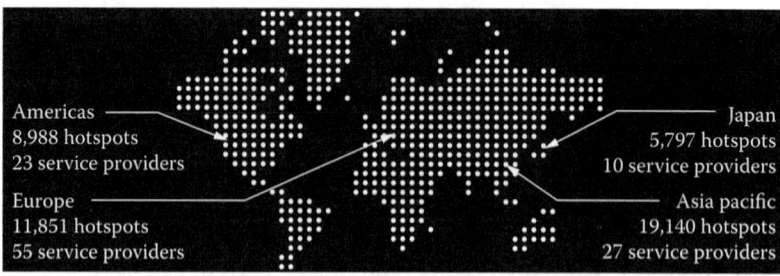

Americas
8,988 hotspots
23 service providers

Japan
5,797 hotspots
10 service providers

Europe
11,851 hotspots
55 service providers

Asia pacific
19,140 hotspots
27 service providers

Figure 1.5   Number of Wi-Fi hot spots globally.

years down the road, UWB holds great promise as the next major technology for high-bandwidth wireless personal area connectivity.

However, with the advent of wireless standards for WLAN and WMAN, deployment of these networks is steadily increasing in enterprises, public hot spots and even within homes. Although widespread effective deployment is still years away, these networking options are open to users now. Top market research firms predict that WiMAX product sales would reach $1 billion to $1.2 billion by 2008. The market for long-range wireless products based on 802.16 and the forthcoming 802.20 standard is also expected to reach $1.5 billion by the same time.

The advent of Wi-Fi technology and hot spots is only beginning to meet this need. Offering portable Internet access, hot spots provide connections to users within a limited range of an access point. Although hot spots extend the reach of the Internet, they still tether users to a fixed location. Meanwhile, many users want mobile access — the ability to retain their high-bandwidth Internet connection even as they freely move about.

Always Best-Connected

Telecommunication is the merging of voice, data (WAN), LAN, video, image, and wireless communications technologies with PC and microelectronic technologies to facilitate communications between people or to deliver entertainment, information, and other services to people. People around the globe are mobile, and they want all their communications to support that mobility. Telecommunication represents a convergence of these technologies into networks and systems that serve people planetwide.

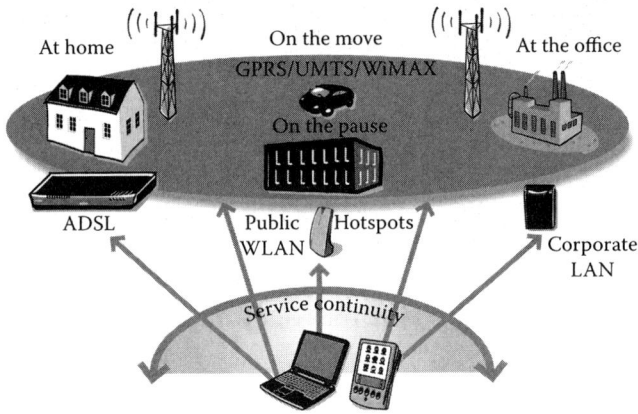

Figure 1.6 Always-best-connected scenario.

As wireless connectivity is becoming more widespread and complex, providing service on the many levels available to wireless users using a variety of devices is also rapidly becoming much more complicated. To accommodate these challenges and to prepare a future in which there are no barriers to access using a handheld, portable, or fixed device, engineers are investigating what measures are needed to create a "Universal Always Best-Connected Communicator," a solution that is capable of communicating regardless of the connection options available to the user.

The Goal

The goal laid out for the telecommunications industry is to keep end users "always best-connected."

To achieve an always-best-connected scenario, users will mix and match mobile platforms and wireless technologies to meet their unique requirements, enabling them to stay connected virtually anytime, anywhere.

Broadband wireless can reach the always-best-connected goal through the following scenario:

- All types of wireless networks will be deployed around the globe.
- Wi-Fi hot spots will proliferate in public places, businesses, and homes.

- Homes and businesses will add UWB (when available) for the fastest distribution of high-definition content.
- First-generation WiMAX technology will be broadly deployed to provide long-distance broadband connectivity for Wi-Fi hot spots, as well as cellular and enterprise backhaul.
- 802.16e WiMAX connectivity to be added in densely populated areas to provide a canopy of wireless broadband data access to mobile laptop users.
- Innovations in 3G technologies will add groundbreaking data capabilities to mobile handset and handheld PC users.

Challenges

Enabling such ubiquitously connected devices poses numerous difficult technology challenges. These include the following:

Multiple radio integration and coordination — Building the handset (or other device) begins with the challenge of integrating multiple radios.

Intelligent networking — This refers to seamless roaming and handoff. Users will expect to roam within and between networks as they do with their cell phone.

Power management — As handsets and other devices evolve to run richer applications, power management will become an even greater challenge.

Support for cross-network identity and authentication — Providing a trusted, efficient, and usage-model-appropriate means of establishing identity is one of the key issues in cross-network connectivity.

Support for rich media types — The addition of a high-bandwidth broadband wireless connection, such as a WLAN or some of the forthcoming UMTS or EVDV/O cellular networks, will open up new opportunities for the delivery of rich media to handheld devices.

Flexible, powerful computing platform: The foundation of a universal communicator-class device must be a flexible, powerful, general-purpose processing platform.

Overall device usability: The final challenge inherent in building a mixed-network device is usability.

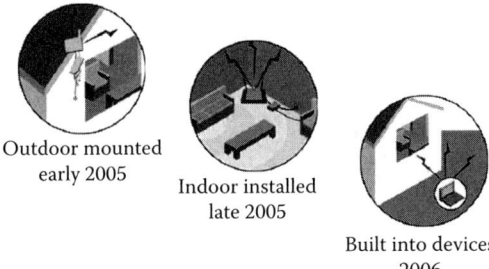

Outdoor mounted
early 2005

Indoor installed
late 2005

Built into devices
2006

Figure 1.7 WiMAX customer premise equipment (CPE) stages.

Key to Success: WiMAX

WiMAX will make ubiquitous high-speed data services a
reality.

Imagine a single wireless technology that can make portable Internet
a reality by extending public WLAN hot spots to metropolitan area
coverage for mobile datacentric service delivery, connecting enterprises
and residential users in urban and suburban environments where
access to copper plant is difficult, bridging the digital divide by
delivering broadband in low-density areas.

Thanks to its innovative technology, WiMAX will offer broadband
wireless access at data rates of multiple Mbps to the end user and
within a range of several kilometers. The same radio technology will
also offer high-speed data services to all nomadic terminals (laptops,
PDAs, etc.) with an optimized trade-off between throughput and coverage. Ultimately, it will enable portable Internet usage, replicating on
the move the same user experience as at home or the office.

Given its huge benefits, WiMAX will develop as a powerful radio
access solution with many integration synergies in mobile or fixed
network architectures. WiMAX will also enable end users to benefit
from an always–best-connected experience when accessing their applications via the best available network, at home, or on the move.

Broadband wireless access has been serving enterprises and operators for years, to the great satisfaction of its users. However, the new
IP-based standard developed by IEEE 802.16 is likely to accelerate

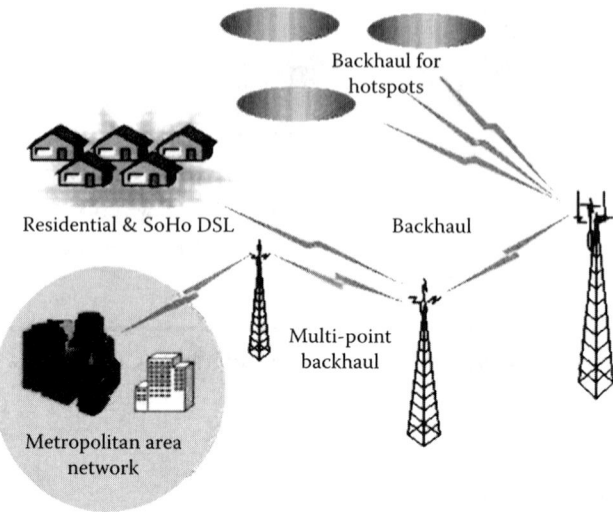

Figure 1.8 WiMAX initial applications.

adoption of the technology. It will expand the scope of usage thanks to the possibility of operating in licensed and unlicensed frequency bands, unique performance under non-line-of-sight (NLOS) conditions, quality-of-service (QoS) awareness, extension to portability, and more.

In parallel, the WiMAX forum, backed by industry leaders, will encourage the widespread adoption of broadband wireless access by establishing a brand for the technology and pushing for interoperability between products.

WiMAX is the right solution to the following:

■ Extending the currently limited coverage of public WLAN (hot spots) to citywide coverage (hot zones), the same technology being usable at home and on the move
■ Blanketing metropolitan areas for mobile datacentric service delivery
■ Offering fixed broadband access in urban and suburban areas where copper quality is poor or unbundling difficult
■ Bridging the digital divide in low-density areas where technical and economic factors make broadband deployment very challenging

Initially, WiMAX will bridge the digital divide, and thanks to competitive equipment prices, the scope of WiMAX deployment will broaden to cover markets in which the low POTS penetration, high DSL unbundling costs, or poor copper quality have acted as a brake on extensive high-speed Internet and voice over broadband.

WiMAX will reach its peak by making portable Internet a reality. When WiMAX chipsets are integrated into laptops and other portable devices, it will provide high-speed data services on the move, extending today's limited coverage of public WLAN to metropolitan areas. Integrated into new-generation networks with seamless roaming between various accesses, it will enable end users to enjoy an always-best-connected experience.

The combination of these capabilities makes WiMAX attractive for a wide variety of people: fixed operators, mobile operators, and wireless ISPs, but also for many vertical markets and local authorities.

Chapter 2

Setting the Stage

A knowledge economy is one that relies intensively on human skills and creativity, the utilization of human intellectual capital supported by life-long learning and adaptation, the creative exploitation of existing knowledge, and extensive creation of new knowledge through research collaboration, and development.

If the whole history of human civilization was compressed into a day, then modern telecommunications (the telegraph in 1837 and all that followed) would only represent the last 30 s. That is how recent all of this development has been. But within that timeframe, much has happened. In that last half a minute, we have achieved a great deal more than what we did during the rest of the day, and progress continues to accelerate.

The technology boom that we are witnessing now would undoubtedly have been science fiction stuff for us in the beginning of the twentieth century. The world entered the twentieth century without planes, radios, or televisions. It enters the twenty-first century with nuclear power, space travel, computers, cell phones, and the wireless Internet. Within the span of 100 years, entirely new fields of science and technology came into existence, and the fundamental political and economic structure of the world changed not once, but several times.

Few would disagree that technology underpins the unprecedented levels of prosperity we enjoy today, with progress in many areas ranging from stem cell research to sending expeditions to Mars. These advances have changed the very premise of our existence.

Although there are many innovative technologies responsible for the constantly evolving current status, information and communication technology (ICT), being in the forefront, is the WMD (not the one talked about a lot in the American and European media) — Worldwide Medium for Development.

ICTs are basically information-handling tools — a varied set of goods, applications, and services that are used to produce, store, process, distribute, and exchange information. They include the old technologies of radio, television, and telephone, and the new technologies of computers, satellite and wireless technology, and the Internet. These different tools are now able to work together and combine to form our networked world, which is a massive infrastructure of interconnected telephone services, standardized computing hardware, the Internet, radio, and television, reaching into every corner of the globe.

In the context of today's competitive business environment and ever-changing social landscape, information has become a critical resource that allows people and organizations to be more productive, effective, and efficient. Irrespective of application area, knowledge is a vital commodity. Information and knowledge assist decision making by enhancing the understanding of problems and expanding the choices based on which decisions are made.

Access to information and knowledge is more widespread today than ever in the past, owing to the popularity of the new but cheaper methods of ICT. Its pervasive effects have positive implications in almost every walk of life, including business, market, education, healthcare, culture, and governance.

ICT has progressed in leaps and bounds since time immemorial, but its growth during the last decade has been spectacular. Entrepreneurs, bureaucrats, social thinkers, and politicians are now advancing views about how individuals, enterprises, societies, and even nations can reap the benefits of this unprecedented technology advance.

The best part about this euphoria is that unlike various other technologies, ICT applications can be tremendously beneficial to a wide spectrum of the population, irrespective of their present economic status. In fact, there are many examples of underdeveloped societies riding the ICT bandwagon and leapfrogging into a knowledge-based economy.

The Knowledge Economy

A knowledge economy is characterized by a culture of innovation and collaboration. Such a culture has some key characteristics such as incentives for innovation, intellectual property protection, and networked constituents. The flow of information and ideas are dominant forces in a knowledge society.

The rapidly increasing ease and speed of information flow is blurring all the boundaries, creating an environment of transparency, collaboration, and value sharing. Today, knowledge needs to flow by connecting the right people, and diverse groups working in different locations, different time zones and, often, having different competencies and skills.

Almost all economic activities are converging into a single space based on the flow of information and ideas, making it a foundation on which a knowledge society is built. This emerging information-flow economy comprises a vast array of stakeholders, and every stakeholder in this convergent space is facing new competitive threats and seeing massive new opportunities open up.

The revolutionary potential of the new ICTs lies in their capacity to instantaneously connect vast networks of individuals and organizations across great geographic distances at very little cost. They have transformed business, markets, and organizations; revolutionized learning and knowledge sharing; empowered citizens and communities; and created significant economic growth for society. ICTs have amplified brain power in much the same way that the nineteenth-century Industrial Revolution amplified muscle power.

A word of caution, though. Access to information and knowledge sharing depends heavily on technological advancement, but its success is largely determined by education, capabilities, resources, transparent societies, capacity to generate and utilize knowledge, connectivity and the availability of diverse content and applications, and the policy and legal or regulatory framework. The key to success in this knowledge economy is a strong foundation, namely, connectivity.

Phases of the Knowledge Economy

We live in a new world, with new dimensions and new horizons, where the boundaries of technology and imagination have been stretched. A new information power is shaping a new geography with new cultures, new markets, new players, and new organizational structures.

Access to information can promote trade, education, employment, health, and wealth. One of the hallmarks of the information society — openness — is a crucial ingredient of democracy and good governance. Information and knowledge are also at the heart of efforts to strengthen tolerance, mutual understanding, and respect for diversity.

These words depict a new and third revolution in the history of humankind: the information age, which leads to a widespread distribution of sources of labor, production, and power all around the world, on any network. The first revolution was agricultural; the second, industrial; and the third, informational, which is symbolized by the coalescence of the worlds of information technology (computers), communications (telephone), and media (television). These three revolutions have been characterized by three different instruments of power: land for the first revolution, capital for the second, and knowledge for the third.

The Island Phase

In this phase, up to, say, the late 1970s, computer, telecom, and broadcasting systems were highly distinctive. Information technology facilities during this early phase of the knowledge economy were few, physically large, and cumbersome, but very low in terms of power when compared to modern systems. Mainframe and minicomputers were used chiefly in very large enterprises and government.

Each computer served a large number of users, but only experts were doing more than just data entry. High levels of expertise were required to operate computers, and the visual displays and keyboard interfaces were very basic.

Public attitudes to the new technology were very mixed. Fears about the dehumanizing effect of large databases coexisted with awe of computers. Government policies typically supported national champions (with their own designs and standards). Organizations concentrated information technology facilities in data processing centers, centralizing information processing.

The Archipelago Phase

This phase, say the 1980s, is characterized by a proliferation of devices of many sizes, usually with limited (two-way) communications. Telecom deregulation and support for strategic research programs on satellite television were introduced in many countries. At the same time, many new industrial and consumer products using microelectronics were widely diffused.

The personal computer found large markets in offices and homes, though early online information systems were (with a few exceptions) disappointing. Public fears about the impact of information technology use on employment were joined by the concern about deskilling, though the workplace trend was more one of work upgradation. Isolated components of the existing division of labor were frequently automated, but there was much less systematic reorganization of work structures and integration of different functions.

The decentralized use of personal computers (mainly as stand-alone devices) caused problems for corporate data processing managers. Equally, economists were puzzled by the lack of reflection of information technology investment in productivity statistics.

The Continent Phase

In the 1990s, the planet was crisscrossed by information superhighways and networks bridging islands of automation. The Internet became a near-universal medium for computer linkages, and mobile systems of many kinds became prominent for voice and data communication. This is not to say that networking was universally diffused — many computer systems remained stand-alone. And the Internet was not particularly easy to use. Many organizations required new skills in the form of network administrators and managers, Web site authors and editors, etc., and its effective use required considerable change in organizational practices.

But as access to the Internet became widespread and the Web provided a design paradigm for information exchange, the online transfer of data mushroomed. Existing services migrated to these media en masse, reaching out to broader and less specialized user bases and exploiting the lessened learning costs of a common interface.

E-commerce applied new information technology to the transactional elements of economic activities and, despite the stock market boom-and-bust frenzy, it does represent significant network integration across the islands of automation of factory floor production, warehouses, offices, etc. It offers scope for new modes of doing business, integration of internal and external processes, and restructuring of supply chains. This requires considerable organizational learning and reengineering.

By the turn of the millennium, there was evidence of performance improvements in information-technology-using firms and new trends in the U.S. economy, suggesting that increasing networking or organizational learning was beginning to overcome the productivity paradox.

The notion that we are now living in a knowledge economy is actually a product of three distinct ideas concerning the role of telecommunications, computers, and information in society.

The first key idea is that largely as a result of developments in telecommunications technology, the world has entered a global era in which our primary frame of reference is — or should be — the world as a whole rather than family, tribe, culture, religion, or nation.

The second key idea is that we are living in an information economy in which the production, processing, and distribution of information has become a very important economic activity in its own right.

The third key idea is that as a result of developments in computer technology, there is no reason in principle why intelligent machines cannot carry out activities previously reserved for human beings and perform them at least as well if not better than their human counterparts.

The Need for Connectivity

> Connectivity is the unbiased transport of information between two endpoints.

Connectivity provides unique opportunities for economic growth and human development. It can shape and enhance a wide range of business, social, and academic applications — from E-commerce to access to financial markets, generating employment to improved agricultural practices, long-distance education to telemedicine, and from environmental management and monitoring to prevention and management of disasters.

Connectivity is not a new concept. After all, it goes back at least as far as when people interacted with each other using symbols and gestures. With the advent of new technologies such as the Internet, mobiles, wireless, and satellites, connectivity has never been the same. Connectivity brings access to information, which unleashes human capital and increases productivity and knowledge sharing, especially in underserved areas where it has been most constrained. Today, connectivity is riding the wave of new forms of communication and is shrinking our world, giving us new ways to interact and, in the process, transforming the way we carry out economic and social activities.

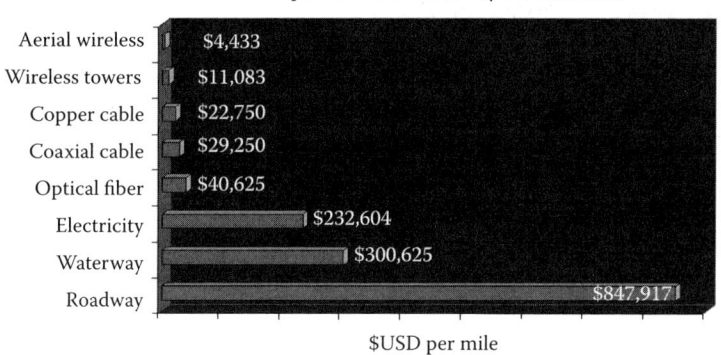

Figure 2.1 **The changing cost of communication.**

As communication between people becomes more fluid and pervasive, it is dramatically changing the structure of society and how people interact. E-mail, short message service (SMS), instant messaging, cell phones, online forums, chat, and videoconferencing — all allow and even encourage ways of communicating and relating with others that are fundamentally different from what has come before.

Digital communication is resulting in a substantial broadening of connections between people across the globe. This enhanced connectivity has benefited both businesses and society alike.

Communication: Evolving with Mankind

The first species of humans resembling us was *Homo erectus*, who not only walked upright but also knew how to make fire and cook food. They traveled over land bridges from Africa and began to populate the world about one million years ago. It is from these first "upright men" that we got our basic body language and the beginnings of speech. But it would be another half a million years or more before they started to draw or write.

The first human carvings date from around 45,000 BC — some Neanderthal man had made scratchings on a mammoth tooth, which was discovered in Hungary. The first drawings of animals date from around 30,000 BC and were found in Germany and France. It took around 30,000 years from the first known drawings or carvings before the appearance of the first organized civilization. This was the Sumerian society in Mesopotamia (modern Iraq), who used tokens for accounts and bookkeeping around 5,000 BC.

The Sumerians had the first known writing, using picture forms. At about the same time, on the other side of the world, the ancient Chinese were also starting to write. By about 1000 BC, the first civilizations had between them developed the first encyclopedia (Syria), alphabets and libraries (Greece), also a postal service and newspapers (China).

The Greeks had also developed early forms of telegraph using drums, beacons, smoke, and mirrors. By 200 BC, the first books were appearing, handwritten on parchment and vellum.

By 1 AD, couriers were carrying mail across the Roman Empire. By the time the Western Empire fell, around 450 AD, the first printing presses were appearing in China. Between the fall of Rome and the Renaissance, Europe fell into the Dark Ages. Most scientific progress was taking place in China and Japan, where paper, printing, and books were being developed. By the year 1035, the Japanese were even recycling paper.

The first paper reached England early in the fourteenth century, and so did the Black Death, which killed half the population and set progress back by a century. It would be a further 100 years or so before the first printing presses arrived; first, Gutenberg in Germany (1450s) and later Caxton in England (1476).

Europe's late discovery of books and printing was the key needed to unlock the gates of thought and progress, as the continent recovered from the Black Death. By the middle 1600s, postal services and newspapers were starting to flourish and, by the end of the century, Isaac Newton had formulated his theories of mass, force, and gravity.

The real scientific explosion took place in the eighteenth century: photochemistry, lithography, metallurgy, and many other sciences made huge progress in what became known as *The Age of Reason*. This tidal wave of progress swept though the nineteenth century with the development of electricity and electromagnetism.

Most people trace the dawn of our modern age to the Industrial Revolution in England in the 1780s. At this time, the fastest any man had traveled was around 12 mph — the speed of a galloping horse.

Within 50 years, Britain had become an industrial society, using steam power, machines, factories, surfaced roads, canals and railways, and with the majority of people living and working in cities. By the 1870s, a similar situation had spread across Europe and in the United States.

The century that had opened with mechanical semaphore as the last word in telecommunications would see the introduction of telegraphy, telephony, facsimile, wireless, cameras, recorded sound, the cinema, the cathode ray tube, and the electric tabulator — a predecessor of the computer. And the relentless acceleration of progress continued through the twentieth century, when mankind reached space and started using satellites as a mode of communication.

The Global Brain

The world entered the new millennium with bright hopes but also deep concerns. The global information society of the twenty-first century is beginning to take shape. New vistas of human development are opening before us, presenting mankind with new possibilities for working, learning, and living in harmony under the guidance of our common ideals.

At the same time, old problems are still with us, reappearing in new and strange forms. Where will future jobs come from? How can we close the widening gap between the information rich and the information poor? What must be done to build a peaceful world based on mutual respect between peoples and tolerance of individual differences?

As we face an unknown future, two things are certain. The first is that the world of communications is being changed by the joining together of broadcasting, telecommunications, and information technology. The second is that this new technological complex — the sensory organs, cerebral cortex, and central nervous system of the information society — will profoundly affect global development and individual destiny.

There are curious parallels between the human brain and human society. Nearly 100 billion deeply connected neurons make up the brain. Each neuron can trigger approximately 1000 other neurons by firing a very-low-voltage electric impulse. In turn, any two neurons in the brain are separated by no more than four or five interconnects. All of our thoughts and behaviors emerge from the interactions between these billions of neurons and are stored as interrelationships between these neurons.

Human society today is gradually moving toward the formation of a global brain that is increasingly becoming similar to the human brain. Soaring connectivity is giving rise to what increasingly resembles a global brain. Connectivity allows the incredibly rich flow of information and ideas that create this single mind and that can integrate all of our intelligence and insight.

The worlds' population is around six billion. The average person in the developed world knows around 300 other people, and the vast majority of people in the world are now connected by less than six steps. This is far more about the number of people who are connected, rather than the connections themselves.

The global brain is very similar to the human brain not only in structure but in the way it functions also. There are two key aspects to the thinking process of the global brain and the individual minds that comprise it. The first is generating and developing ideas or, in other words, information creation. The second is filtering the universe of information, paying attention only to what is important and useful or, in other words, information-based decision making.

Ideas are what help us face the conditions and challenges confronting us. We face different conditions and challenges daily, or more appropriately, at every moment. Human beings cultivate idea formation by experience or induced skills and knowledge. The global brain also generates ideas in the form of information.

The information that assails us, effectively filtered by our brain, is essential for our survival. We would be completely overwhelmed if we were not able to reduce the millions of sensory impressions we receive to something our logical brain can cope with. Schizophrenics can be understood as lacking the usual filters that would protect them from being swamped by their sensory input. Instead of perceiving only the outstanding features of their environment, everything stands out for them.

In the information age, this ability to filter effectively has moved from an essential of survival to one of the primary determinants of success. Information overload is the defining feature of our times. Those who are most effective at making sense of the flood of incoming information and turning it to action lead our world. Filtering performed at the level of the global brain is called *collaborative filtering*. Instead of everyone individually attempting to make sense of the universe of information we swim in, we can work together.

Now connectivity is extending our senses to all the connected people on this planet. However, just a small proportion of the planet's population is connected. It is critical that we extend participation as broadly as we can. Efforts to achieve universal connectivity, particularly at the lower-income levels, in all countries, and especially in developing countries, will require innovative approaches and partnerships, including group and community connectivity and private sector investment. In this regard, the establishment of integrated multipurpose and multimedia community information centers will be important.

The Digital Divide: What Does It Signify?

The swift emergence of a global information society is changing the way people live, learn, work, and relate to each other. An explosion in the free flow of information and ideas has brought knowledge and its myriad applications to millions of people across the globe, creating new choices and opportunities in some of the most vital realms of human endeavor. Yet, too many people, especially from the not-so-fortunate economies, remain untouched by this revolution. A digital divide threatens to exacerbate the already wide gaps between rich and poor, within and among countries. The stakes are high indeed.

The concept of the digital divide denotes the gap in access to information resources in some countries compared to those with state-of-the-art networks: telephone, radio, TV, Internet, satellite, in short, anything that can be classed as ICT. Thus, the digital divide refers to the difference in facilities for people to communicate, relative to their geographic location, living standard, and level of education. Ultimately, it is an indicator of a country's economic and social situation.

Modern societies are currently undergoing a number of fundamental transformations caused by the growing impact of the new ICTs on all aspects of human life. But this revolution, brought about by the new technologies, has to confront a major challenge, namely, the extreme disparities of access between the industrialized countries and the developing countries and those in transition, as well as within societies themselves. Even though there has been a substantial increase in telecom investment, not to forget technological advances in the past decade, there are still enormous gaps between the developed and developing world in accessibility to telecom, and within the developing world, between urban and rural areas.

There is still an average teledensity in decimals in the poorest countries, whereas in some advanced countries it is touching saturation levels. The gaps are even greater between urban and nonurban areas. There are almost three times as many telephone lines per 1000 in the largest city of lower-middle-income countries as in their rural areas, and more than seven times as many lines per 1000 in the largest city of low-income countries as in their rural areas. These gaps are even

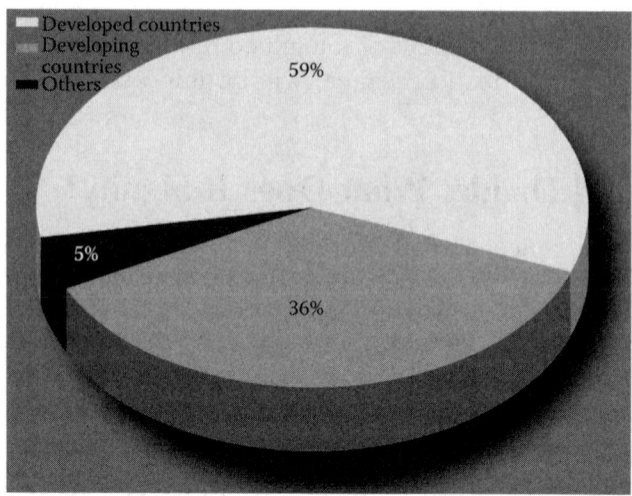

Figure 2.2   The digital divide — Internet Users 2004, ITU, and UNCTAD.

more significant, given that more than 50 percent of the population and as many as 80 percent in the poorest countries live in rural areas.

Indeed, the real issue is how to take account of the human dimension of the digital divide between and within countries. Despite increased awareness, the rich–poor divide in economic well-being is growing. The challenge now lies in enlisting technology as an ally in the movement for development and social equity. Affordable access, connectivity, and the skills to utilize increasingly advanced but essential services remain the central public interest issues in the area of ICTs across the globe. This is true for all countries, but particularly for developing countries (Figure 2.2).

In many developing countries, less than one percent of the population has Internet access. However, it is clear that the numbers are small and the distribution limited. Although it is fully understood that for most of the more than one billion people in these countries who earn less than $1 a day, food and clean water are probably higher priorities than Internet access. However, beyond the absolute basics, access to the networks will be critical in helping them improve their lives and seize the opportunities that lie ahead.

At a time when information is power, the inequities of access to and dissemination of information extend to citizens' differential ability to be politically or economically effective. The contemporary policy dilemma is that the urban centers are connected to global networks, illuminating critical paths of planetary contact and influence, whereas rural areas languish in isolation, and the gap between those that are connected and those that are not is widening.

There are many reasons behind the polarization of today's knowledge society on the basis of access to connectivity, and hence information. Some of the vital issues responsible for the digital divide across the globe are lack of resources, scarce infrastructure, widespread illiteracy, inadequate technology, biased policies, apathetic governance, political instability, and deep-rooted corruption.

Various studies and surveys conducted in the past illustrate one surprising aspect regarding the causes of the digital divide. Technology, though considered undeniably important in comparison to other causes, is rated less important than policy, funding, private sector participation, and foreign cooperation. This point was illustrated at the ITU Telecom Africa 2001 Policy Development Forum, where in response to the question: "What is the most important barrier to the provision of access to all Africans?" participants ranked lack of funding (47 percent), regulation (23 percent), lack of public and private sector cooperation (18 percent), and inadequate technology (12 percent) as the barriers.

All these issues are interrelated and have technology as an insignificant component. But in the recent past, advances in information and communication technology have had a revolutionary impact on these obstacles, albeit indirectly. These radical developments were based on a wave of concurrent technological innovations (in informatics on one hand and in telecommunications on the other), underpinned by a number of external factors (network externalities, knowledge-sharing effects, and innovative business modeling) never experienced in the past.

Bridging the Digital Divide

It may seem quite intriguing to believe that ICT tools such as the Internet, computer, and even telephone can contribute to the local development of communities that are often disadvantaged by the lack of even more basic facilities, such as drinking water, roads, or electricity. In this situation, ICT investment may not even look economically realistic, let alone being useful or a priority.

But various case studies show that it is possible to develop innovative services and solutions around ICT that meet the basic everyday needs of the local economic and social organizations and the poorest people. ICT could contribute to a lasting, integrated development

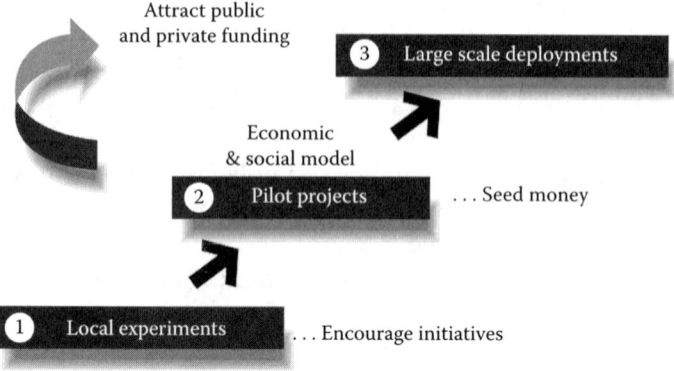

Figure 2.3   Process dynamics — bridging the digital divide.

process by offering local high-value-added proximity services that take into account the people's way of life, real needs, and incomes.

To remote and isolated communities that lack even basic infrastructure such as roads and electricity, the ability to access information to take care of oneself, feed oneself, communicate with one's peers, develop one's own projects, and so on appears like a breath of fresh air. Realistic investment in Internet or other ICTs operated by a trained local community, coupled with an innovative business model, may never replace the roads that are so sadly lacking, but will make it possible to ensure better use of what few means are available (Figure 2.3).

Some of the common characteristics emerging out of various successful initiatives of this kind show that both local players and local residents can achieve a genuine "leap forward" — economically, politically, and socially. These characteristics are based on two converging virtuous circles, one impacting economic activities and the second impacting social and political establishments.

Such services can reduce logistic, marketing, and distribution cost for communities of fishermen, farmers, and other rural producers. For example, ICT can help create local, more transparent marketing channels, thus limiting speculation and the risk of artificial shortages and improving the distribution of margins between the various links in the value chain of each sector, from producer to consumer. Time and money saved in this manner can be ploughed back into productive new activities, helping boost the local economy and leading to the creation of jobs. This will, in turn, justify more communication resources, and so on. This is the first virtuous circle.

ICT can be used as a tool to support the implementation of health program initiatives in which information campaigns are very important. In the areas of education and administration, ICT has the potential to improve communication between public authorities and local people, as well as between central and local authorities. It will facilitate greater transparency in how institutions are run, thus moving toward the objective of good governance. This is the second virtuous circle.

Wireless Broadband: Connecting the Poor

The promises of wireless broadband Internet technologies have generated much interest on the part of the international development community. Whereas in developed nations these technologies have primarily been associated with mobility applications and local area networking in homes and offices, their most intriguing application in developing nations is the deployment of low cost broadband Internet infrastructure and last mile distribution.

The rationale for such interest is simple in theory: The digital divide cannot be resolved any time soon because of the prohibitive cost of deploying conventional wired infrastructure in developing countries. However, wireless broadband Internet is a very effective and inexpensive connectivity tool that has the potential to solve this bottleneck, and it can be the most promising accelerator of technology adoption in developing nations. International development experts and leading IT corporations also consider wireless Internet technologies essential to bridging the digital divide in developing countries at a manageable cost and within a reasonable timeframe.

Wireless broadband Internet, though central to the developing world that is determined to forge ahead with information society and knowledge economy initiatives, cannot work any magic by itself. It can only be successfully deployed as demand for connectivity and bandwidth emerges in support of relevant applications for the populations served. These applications can be E-government, E-education, E-health, E-business, or E-agriculture applications. In other words, wireless technologies can only facilitate development, but it is innovative ICT applications that drive development.

The Digital Enterprise: Changing the Way Business Is Run

The digital revolution has the potential to spread access to knowledge, information, and markets to enterprises that have traditionally been excluded from these crucial aspects of trade. The power of connectivity has dynamically changed the business landscape because of the wealth of resources it has made available to those who are connected and have the skills to tap this abundant resource.

Technology has rapidly emerged as the central force in business today. As a conduit for business transformation, the technology has altered long-standing business models, relationships, processes, and infrastructure. Modern enterprises, as they respond to these changes, face unique opportunities and obstacles arising not only from the pace of change but also from the very nature of the change itself. Technology has generated extraordinary opportunities both for new business ideas as well as for new ways of implementing old ideas. To become a successful player in the present business environment, enterprises must continuously analyze existing market conditions to identify emerging opportunities, and they must then move quickly to embrace viable opportunities.

The adoption of technology by enterprises has grown considerably over the past few years, with more and more firms getting hooked. Firms use technology mostly for internal automation, for example, of office and production processes, for customer relations and supply-chain management, for the management of distribution and logistics networks, for business information and analysis, and last but not least, to communicate internally as well as externally. The arrival of hyper-connectivity and living networks has implications for almost every aspect of business. Business is being transformed faster than ever. Already, relatively recent technologies such as e-mail, mobile telephony, and text messaging are changing the way people communicate and the way companies work.

Outsourcing, which is one of the hottest and most extensively debated phenomena today, can be traced back to the 1980s, when its seeds were sown in form of remote services or teleservices.

The Case of the Teleservices Industry

In the 1980s, businesses in the United States and Western Europe began to realize the potential benefits that telecommunications could offer in terms of wider customer access and improved service care. A highly competitive business climate, falling telephony costs, and high telephone penetration combined with consumer demand helped create a boom in teleservices employment. In the United States, call centers expanded quietly in the 1980s, aided by the telecom companies' decision to offer 1-800 toll-free calls from any region within the country. Today, call centers employ close to four million Americans.

The rapid growth of call centers that occurred in the 1990s in the United States began to level off after the year 2000. These leveling numbers can be explained, in large part, by the offshoring of these teleservice jobs. Initial estimates show that more than 1.9 million service jobs have been exported from the United States since 1995. The Forrester Research Group estimates that 3.3 million service jobs will leave the country by 2015, a number that is considered conservative by many analysts. Economists from the University of California at Berkeley estimate that the total number of U.S. jobs vulnerable to being outsourced is around 14.2 million.

Whereas the costs and benefits of outsourcing to developed countries are hotly debated in North America, developing countries are eagerly grabbing these jobs. India, with its 25 million well-educated English speakers, has been a major beneficiary. The typical call center agent in India is a young, recent university graduate working on a full-time contract. An equal number of men and women are employed in this sector and when offered jobs, the overwhelming majority of applicants accept.

In the services trade, many highly skilled tasks are already being performed in developing countries. Engineering, litigation, design, and investing services are all currently being imported from less-developed countries (LDCs). In the teleservices industry, jobs that can be standardized and are rule based are the easiest to transfer, and this is where most of the early growth has been. Companies are effectively using call centers to save up to 60 percent off original home-country costs. These phenomenal savings are driving companies worldwide to continue this trend, constantly pushing its limits.

As businesses and consumers warm up to this phenomenon, we will continue to see everyday services transferred abroad. The limits of this type of trade are restricted only by the imagination and the creative abilities of entrepreneurs worldwide. The economic forces and the technology enabling these practices are growing stronger and penetrating deeper into society every year. As this process unfolds, the capabilities and opportunities for employment growth in teleservices for both LDCs and developed countries will grow.

Teleservices have the potential to be both a positive and significant tool for social and economic development worldwide. Today we see remarkable examples of people providing high-skilled services over fiber-optic cables and across continents. Growing wage differentials between regions are driving these phenomena well into the future. Ultimately, it is up to the citizens and entrepreneurs of developing countries to decide to what extent and to what level they can service the global market. It takes a deep understanding of innovative technologies and an imaginative mind to create new business ideas and transform them into reality. Fortunately, many people around the world are doing just that, and it is this entrepreneurial spirit that has emerged as the engine of global economic growth.

Digital Market: Impact of E-Commerce

> Electronic commerce can be defined as "the buying and selling of information, products, and services via computer networks." The definition could be extended by including "support for any kind of business transactions over a digital infrastructure."

Internet and E-commerce have indeed changed everything. Beyond the hype and the headlines exists the real work of building new business structures and technologies that can ensure success in a rapidly changing business world. The compelling reasons to bring businesses online apply to just about any organization, whether in the manufacturing, distribution, merchant, service, or any vertical industry marketplace.

Foremost among these, the growth of E-commerce has been explosive. E-commerce is more than online transactions between buyers and sellers. It is the only way to compete in today's changing business environment. The real power of the technology comes from improved business efficiency and customer service. Effective E-commerce solutions focus on the complete sales process, i.e., marketing, sales, customer support, and communication with suppliers.

By integrating proven business applications, systems, and data with the rich multimedia functionality of the Internet, the entire operation can be streamlined while building a solid customer base and driving sales. E-commerce offers unlimited opportunities to leverage the Web's global reach and generate new revenues by tailoring the business model to the Web.

E-commerce provides new ways to reach wider markets, enhance service locally, and accommodate seasonal sales cycles. It also presents opportunities to complement existing channels and relationships while reducing business cycle times, improving cash flows, reducing inventories, decreasing administrative costs, and opening new marketing and sales channels. Further, E-commerce can promote products and services to a global market and expand sales without investing in bricks-and-mortar storefronts worldwide.

E-commerce is a vital tool for providing self-service opportunities. It thus delivers high-quality, low-cost customer support to a larger number of customers without increasing the support staff in proportion, which can result in higher profits. Providing after-sales customer support can transform customer satisfaction into customer loyalty. This decreases the cost to serve each customer, and increases convenience for new and existing customers.

Digital Government: Assessing E-Government

> E-government is the use of ICT to promote more efficient and cost-effective government, facilitate more convenient government services, allow greater public access to information, and make government more accountable to citizens.

E-governance is not just about government Web sites and e-mail. It is not just about service delivery over the Internet or digital access to government information or electronic payments. It will change how citizens relate to governments as much as it will change how citizens relate to each other.

E-governance is much more than information technology. In this wired-up era, the inhabitants of knowledge societies will have all the more freedom, flexibility, and opportunities to decide how they would like to be governed and by whom. The underlying truth will become even more self-evident — it is not the leaders who govern people, but it is the people who let the leaders govern them.

It will bring forth new concepts of citizenship, both in terms of needs and responsibilities. E-governance will allow citizens to communicate with government, participate in the governments' policy making, and allow citizens to communicate with each other. E-governance will truly allow citizens to participate in the government decision-making process, which will thus reflect their true needs and welfare.

E-governance presents challenges and opportunities to transform both the mechanics of government, and the nature of governance itself. It affects all government functions and agencies, the private sector, and civil society. Over time, it has the potential to change the way government operates, and how citizens and businesses interact with government.

Most believe that the person who will transform the present government into E-government would be an IT professional. But a successful E-government is a result of joint cooperative effort of politicians, bureaucrats, employees, industry, and IT professionals.

With the emergence of proactive knowledge societies, governments will have no choice but to constantly improvise to bring in greater efficiency, accountability, and transparency in their functioning. People are becoming more aware of their rights and the opportunities that lie ahead and are developing capabilities to make informed choices in all areas that influence them, including the sphere of governance.

Transparency and free flow of information are the minimum conditions for achieving good governance. For good governance, it was not enough to have merely a democratic constitution but also a culture of respect for human rights and dignity. Accountability and transparency are essential for any government that wants to serve its people. Thanks to IT, the era of E-governance is a reality today. That means the old system and procedures of governance have become less important.

Just as the Industrial Revolution did not end agriculture, so also the Information Revolution will not end government and governance. As the usefulness of information, IT, and information work increases, E-governance will find more ways to substitute them for the old style and methods of governance, expensive and oversized administration, big office buildings, and hordes of government employees. IT will lead to administrative and management revolution. The file-pushing processes will be done away with and data processing and decision making will be done quickly and cheaply.

Achieving E-governance will be challenging. The big challenges are not technological but cultural. Achieving E-governance will require a change management process that builds awareness, understanding, trust, common purpose, and a genuine willingness to change if it is to move from an idea to reality. Most of the thousands of government Web sites around the world are simple informational sites. Very few E-government portals offer transactional functionality. The real E-governance models will be where the back-office data processing gets linked with the front-end Web interface of citizens. E-governance is the future, and we must go in for it to make the future secure for future generations.

The Economic Impact of Telecommunication

The impact of Christopher Columbus' discovery of the New World was immense. It opened up a new era of trade between Europe, the Americas and, ultimately, the rest of the world. Vast sources of new wealth were discovered; new forms of economic, social, and political organization were invented; and new ideas and new forms of cultural expression were created. At the same time, a high price was paid. Cultures were destroyed; people were annihilated, dislocated, and cast into slavery; and the ancient wisdom was lost. Similarly, telecommunications, 500 years later, is again leading mankind into an unprecedented type of civilization.

Telecommunication, having played a crucial role in making developed economies more productive, will be a critical player in the development of economies around the globe in the twenty-first century. It provides the infrastructure and services that are driving the knowledge economy, and it can enable other sectors of the economy to move toward sustainability.

ICT can make a dramatic contribution to achieving sustainable social and economic development goals. By transforming communication and access to information, the telecommunications industry can create powerful social and economic networks that can underpin sustainable development in emerging economies. The rise of the global information economy has transformed human life — nationally, regionally, locally, and within the family. Today, everything is changing because of telecommunication: be it the nature of work, relationships between people, media, messages, and patterns of political life.

The role of the telecommunications industry as an enabler of greater sustainability in other industries is very significant. All segments of the economy, i.e., industry, commerce, society, governance, and agriculture, benefit by the intelligent use of telecommunications products and services. Innovative use of telecommunications technologies has generated untold numbers of opportunities for businesses old and new, big and small. These opportunities include but are not limited to the following:

■ A tremendous new way to reach customers and gain advantage over competitors
■ Instant global reach
■ Streamlined distribution channels and logistical operations
■ The ability to cut the price of doing business, lower procurement costs, and drive costs toward zero

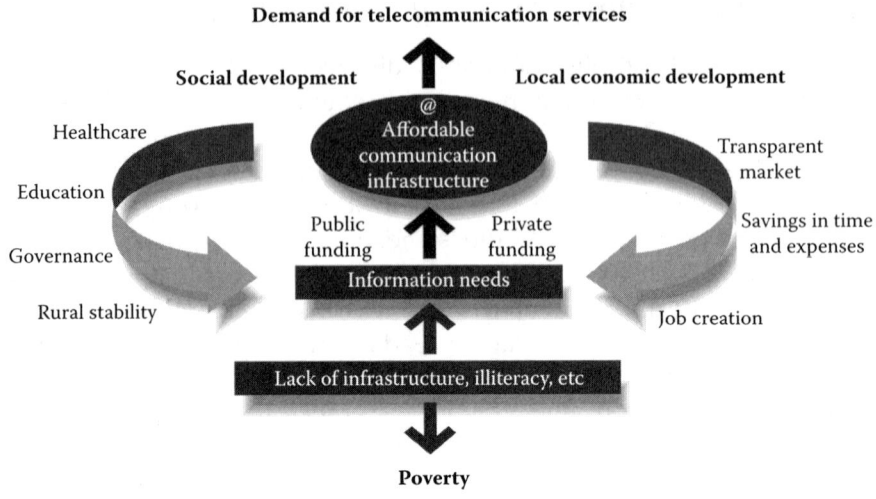

Figure 2.4 Demand for telecommunication and poverty alleviation.

Telecommunications technologies also indirectly enhance productivity in modern operations by reducing their use of resources through many methods such as smart energy management, more efficient transport, transport substitution, dematerialization, E-commerce, and substitution of services for products.

Telecommunication has created a new wave of business and economics, with great profit maximization. For instance, in 1994, when the telecom revolution experienced an upsurge, the 10 largest telecom giants made bigger profits than the 25 largest commercial banks.

The developments in telecommunication impact on various spheres of human activities: the socioeconomic decisions that people make, concepts of national borders, patterns of international trade, and so forth. Telecommunications issues have become items on national, economic, and social development agendas. More and more people are gaining access to telecommunications services ranging from basic telephony to various value-added services.

So far in this chapter, we have discussed the concept of the knowledge economy, the scope of ICTs, issues of information access and connectivity, impact of information revolution on society, business and governance, and economics of telecommunication. Telecommunications infrastructures are the major forces that drive developments in all of these issues.

Developments in the telecommunication sector are accelerating at a spectacular rate. Investments are escalating, and multinational telecom corporations are expanding their activities globally. To understand the impact that telecommunications technologies are having on our society and what to expect in future, we first need to understand the telecommunications industry. The next chapter provides an overview of the telecommunications terrain and presents the concepts and issues related to the subject.

Chapter 3

Telecommunication: A Connecting Mechanism

The rapid development in communications will change the world. In simple terms, telecommunication can be defined as the process of communicating information via electronic means over a distance.

The telecommunications industry has undergone a phenomenal transformation in recent years. Ever since its inception more than a century ago, the worldwide telecom network has enabled universal interpersonal voice communication. The exact modalities under which this service has been provided have evolved over time, often because of technical changes: from connection through a human operator to electromechanical to electronic switches, from only local or national calls to seamless international calls, from fixed-line access to wireless access, from one basic service to a myriad of value-added services.

Technological progress has brought about mobile communications. The Internet has revolutionized business life. New services and escalating transmission speeds have opened up novel applications. Privatization and deregulation have created widespread competition. Institutional reforms have resulted in ever-greater contributions to national economies. Requirements of individuals and businesses have become increasingly sophisticated. All of these have lowered the barriers of time and space (Figure 3.1).

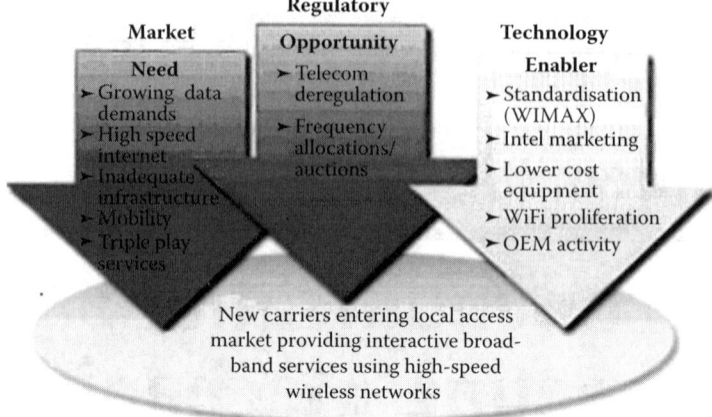

Figure 3.1 Disruptive forces acting on telecom landscape.

The most significant development in the communications industry in the past ten years has been the dramatic rise in network capabilities and the subsequent fall in communications pricing. These forces have opened the floodgates for trade on a global scale. The effects of this revolution have been felt in almost every sector including migration, investment, real estate, education, trading, governance, and law. The empowering capabilities of advanced communications technologies will certainly be the pivotal force shaping economies and societies over the next 50 years.

Telecommunication: Continuously Evolving

Almost 5000 years ago, our ancestors relied on smoke signals for visual (or optical) transmission systems, establishing one of the oldest forms of communication in recorded history.

The next development, sometime around 300 BC, was the use of carrier pigeons to deliver handwritten messages. Beyond that, the only other means of communication was to physically deliver messages to friends, family, and associates located afar.

It was not until 1792 that a new form of optical telecommunication came along: the semaphore. Developed by the Frenchman Claude Chappe, these windmill-like structures enabled people to relay messages at distances of up to 20 mi. However, as with many of today's connections, bandwidth was an issue because the semaphore could

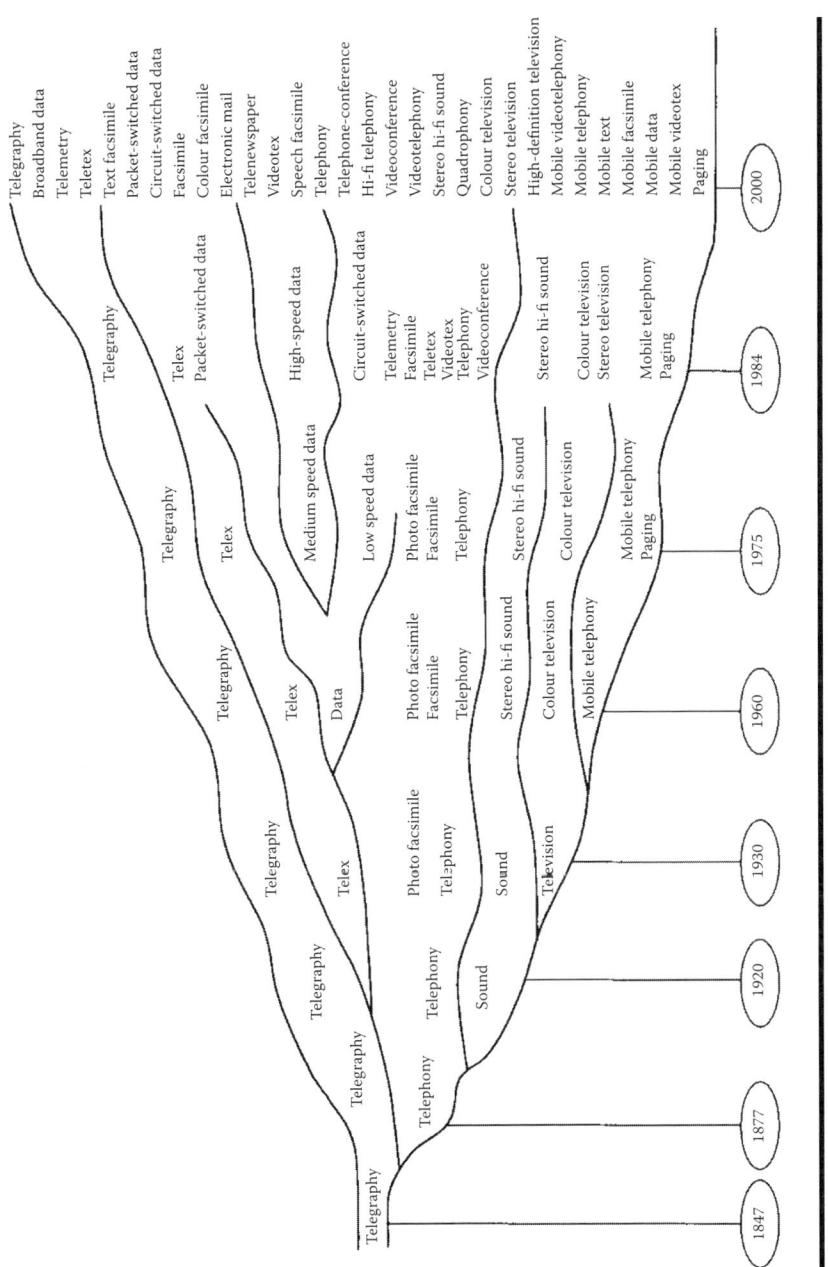

Figure 3.2   Evolution of telecommunication.

transmit only 15 characters per minute. From the invention of the telegraph in 1837, the telephone in 1877, and wireless radio technologies in 1895, communication proliferation has set the stage for modern communications as we know it.

The telephone was invented in 1876 and the telephone exchange in 1877–1878. Soon, the public telephone operators (PTOs) dominated the field by means of national monopolies, and national telephone networks grew in size and number of subscriptions. National monopolies facilitated international cooperation and made it easier for national PTOs to internetwork to offer international telephony services to the public. The fact that telephony services were provided by national monopolies influenced the operational principles of telecommunication. Operators naturally preferred centralized network control. This enabled them to offer reliable transmission and to maintain the integrity of the network. The monopoly situation facilitated this. A hierarchical architecture was soon imposed on the network because centralized control otherwise would have become an unmanageable task as telephony networks grew. The technological implementation of the telephony services was based on chosen design principles. The terminals (i.e., phone sets) and the telecom service provision (i.e., voice transmission) are dedicated to one purpose only (i.e., speech interaction).

The phenomenal technological advances of the nineteenth century brought profound changes, many of which were made possible by the introduction of mechanically generated electricity in 1832. With the ability to harness electron flow through copper wire, European inventors Wheatstone and Cooke, and almost simultaneously, Morse and Vail in the United States, developed the telegraph, and the telecommunications industry was born.

In the late 1960s and early 1970s, users on different networks could not share information and resources. Other telecommunications services, such as telex, telefax, and early computer communication, as well as first- and second-generation mobile telephony, were introduced later. It is important to note that operational and technical principles for telephony, as outlined earlier, were passed on to these newer areas of telecommunications. Thus, telephony is at the heart of the telecommunications paradigm.

The success of the computer industry and computer networking was a driving force in the process of digitizing telephone networks. Several groups began developing the concept of internetworking, which allowed computers on different networks to connect and exchange information. Computing and digital technologies were deployed in switches and other network equipment. Later, digital encoding of speech was standardized.

Developments in telecommunications have seen a gradual transition from analog to digital systems. During the 1990s many national telephone networks were fully digitized. Although networks were no longer based on analog technology and the provisioning of dedicated services, many of the operational and technological principles remain, and they still form an important part of the telecommunications framework. The advances in computing led to new opportunities and new demands, for instance, using computers in switches.

In the past, telecommunication was considered a luxury by many governments and development planners, especially in developing countries. They believed that extending telecommunications networks to rural and remote areas (where most of the developing countries' population lives) was too expensive. Today, innovations in satellite and wireless telephony, coupled with solid-state components for digital switching and end-user equipment, have spectacularly lowered the costs of providing telecommunications facilities to any location, from the buzzing city centers to rural villages.

In fact, the growth of telecommunication, especially computer networks, has been the strongest contributor to the globalization and development we are experiencing today (Figure 3.3).

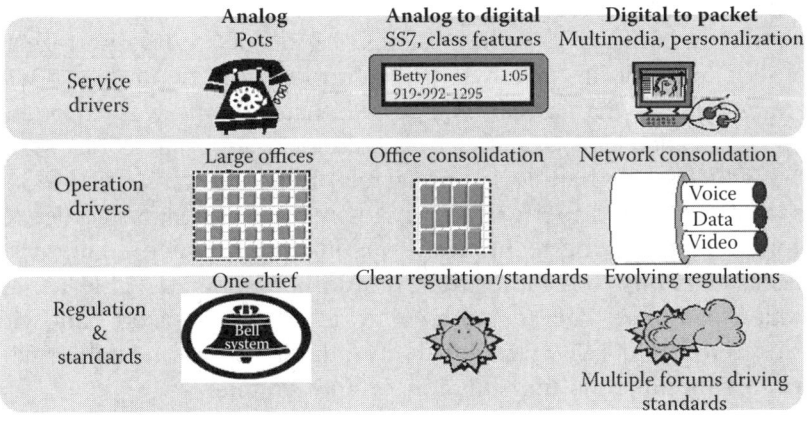

Figure 3.3   Telecommunications network transitions.

Trends in Transmission Technology

Beginning with the telegraph in the 1840s, electronic data communication has greatly speeded up the transmission rate of information. What took days or weeks to transmit during the 1700s could be transmitted in minutes or hours by the 1900s. Today, telecommunications networks transmit huge quantities of information in a fraction of a second.

Almost immediately after the invention of the telegraph in 1837, people saw the need for undersea cables. The first was a cable underneath the Thames in London in 1840. Ten years later, the Brett brothers successfully laid a cable across the English Channel connecting England and France, but it broke within hours when it was snagged by a fisherman. A year later, an armored cable was installed, enabling the commencement of telegraph service between London and Paris in 1851.

Early attempts to lay cable across the Atlantic illustrate the considerable obstacles inherent in installing and maintaining an undersea system. After several years of perseverance, an undersea cable was successfully installed between Ireland and Newfoundland in 1858. Unfortunately, success was short-lived. Not only was the system slow, having a bandwidth of just two words per minute, but it also failed after a few months of operation.

It took the world's largest ship, the *Great Eastern*, with a crew of some 500 men, almost two years to successfully lay the next cable between the continents. The *Great Eastern* returned to the North Atlantic, recovered a broken cable from a prior attempt, and added a second connection in the summer of 1867, giving the world its first modern ring-configured cable systems. As occurs today, performance improved dramatically: bandwidth quadrupled from two to eight words per minute in less than a year. The growth that followed was exponential.

In 1955 and 1956, the first transatlantic telephone cable system, TAT-1, which could handle 48 analog telephone circuits simultaneously, was deployed, and it performed flawlessly for more than two decades.

In 1962, SD analog technology was invented, which enabled bidirectional transmissions rather than requiring a separate cable for each direction of traffic. This technology, which also increased bandwidth, was used to build TAT-3, the third transatlantic telephone cable, which allowed simultaneous transmission of 148 circuits.

Meanwhile, Bell Labs' researchers began to explore the possibility of using light waves and new digital technologies to transmit voice and data. The first optical system was installed in 1982 in the Canary Islands and, in 1988, across the Atlantic. That was a revolution in terms of opening up a whole new medium to the international communications world.

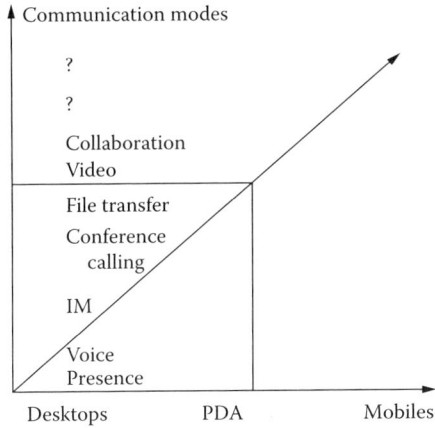

Figure 3.4   Evolution of access device platforms and communication modes.

In 1991, the first successful transoceanic fiber-optic transmission using laser-pulsed light with a wavelength of 1.5 μm was conducted. This development opened the door to higher capacities by increasing the usable bandwidth.

The year 1998 represents another important milestone, when AC-1, the first fiber-optic network specifically designed for DWDM transmission, was commissioned. Spanning the Atlantic from New York to Great Britain, Germany, and the Netherlands, AC-1 had an initial bandwidth of 40 Gbps, spread across four fiber pairs. In another demonstration of engineering expertise, AC-1's capacity was soon doubled to 80 Gpbs.

As transmission speeds and quality continuously improve, one more area of technology development is opened up. The capacity and quality of a network is as good or as bad as its weakest link. For years now there have been many barriers as well as excellent developments related to last mile solutions. Still, we do not know the winner in this area, but we are quite close to the solution, and WiMAX seems to be one of them. To take our discussion further, let us understand the fundamentals of subscriber access technology.

Trends in Subscriber Access Technology

The current goals of subscriber access are to achieve higher speed and greater diversification. The Internet is regarded as the driving force for higher speeds. The use of Integrated Services Digital Network

(ISDN) to access the Internet more quickly is gaining precedence over analog modems, because ISDN can support a greater amount of data (up to 128 Kbps) using existing telephone lines.

However, new access technology will be necessary to implement higher access speeds to meet the future requirements expected for multimedia applications. To address this challenge, two technology areas are being developed extensively: the first is fiber-optic cables, and the other is wireless technologies.

The initial last mile technology was based on copper wires and cables. Copper wire is one of the oldest transmission channels currently in use today. This system is basically used for voice transmission processes. It consists of a pair of twisted insulated wires, and hence the name twisted-pair wire. Coaxial cable provides a higher capacity than the copper or twisted-pair cables. Coaxial cables consist of two wires: The first, a copper wire, is surrounded by an insulator and the second is surrounded by a metallic cylinder called the shield. This design provides the coaxial cable with a special advantage: electrical interference is reduced because the two conductors are shielded and confined separately. The coaxial cable has a greater capacity than the copper wire and has the potential to carry television signals also.

Ease of use and deployment, both of which lower cost, are the major reasons why most developing countries of the world use copper wire for telecommunications transmission for the last mile. Despite this affordability, copper wire has numerous disadvantages: the cost of laying copper cables is high and so is the maintenance cost; further, it is susceptible to corrosion, rain, and theft.

Next came fiber optics, which utilizes thin strands of glass fiber through which light waves travel. These thin strands of glass carry pulses of light rather than electric signals and, as a result of this, they are not susceptible to the electromagnetic interference that is common to most electrical systems. But because of their small diameter, they are difficult to handle and they also create problems in installation and maintenance. Fiber-optic communication has many advantages over "over-the-air-transmission" and the standard coaxial communication system. Fiber-optic cable is particularly useful when interference-free communication is necessary, and a single fiber-optic cable has a large channel capacity and therefore permits multiple uses. Fiber-optic cable offers numerous advantages over copper and coaxial cables: it provides a higher transmission capacity, broad bandwidth, is easily transportable, is immune to electromagnetic interference, and it provides capacity to transmit all forms of communication (voice, data, and video).

Because of the advantages that fiber-optic cable offers (this system is faster, secure, and more interactive than other cable systems), tele-communications companies all around the world are replacing their cable system with fiber optics. But again, for last mile access, it is not yet viable, as it suffers from similar deployment problems as copper (digging trenches, etc.). The second key issue is the need for special devices for termination and connections, which makes it less attractive for the last mile.

Last in the series was wireless technology that used radio for access. The first commercially available radio and telephone system, known as Improved Mobile Telephone Service (IMTS), was put into service in 1946. This system was quite unsophisticated, but then solid-state electronics did not exist at that time. With IMTS, a tall transmitter tower was erected near the center of a metropolitan area. Several assigned channels were transmitted and received from the antenna channels and complete a call. Unfortunately, the number of channels made available did not even come close to satisfying the need. To make matters worse, as the metropolitan area grew, more power was applied to the transmitter or receiver, and despite the reach being made greater, more erstwhile subscribers were unable to get a dial tone.

The solution to this problem was cellular radio. Metropolitan areas were divided into cells of no more than a few miles in diameter, each cell operating on a set of frequencies (send and receive) that differed from those of the adjacent cells. Because the power of the transmitter in a particular cell was kept at a level just high enough to serve that cell, these same sets of frequencies could be used at several places within the metropolitan area.

Two characteristics of cellular systems were important to their usefulness. First, the systems controlled handoff. As subscribers drove out of one cell and into another, their automobile radios, in conjunction with sophisticated electronic equipment at the cell sites (also known as *base stations*) and the telephone switching offices (also known as *mobile telephone switching office* [MTSO]), transferred from one frequency set to another with no audible pause. Second, systems were also designed to locate particular subscribers by paging them in each of the cells. When the vehicle in which a paged subscriber was traveling was located, the equipment assigned sets of frequencies to it, and conversation could begin.

Advances in radio technology, along with growth in demand of wireless technology owing to its inherent features, created more excitement in wireless technology research, which resulted in development of many new wireless technologies such as GSM, CDMA, Wi-Fi and, most recently, WiMAX. Easy deployment (no need to dig trenches), low maintenance, and recent technology developments (which we will discuss in detail later) make wireless last mile solutions very attractive.

Trends in Transmission Coding Technology

Transmission in the telecommunications networks of today is becoming increasingly digital in nature. The term digital, however, does no more than imply a string of ones and zeros racing through the network. But how are these ones and zeros to be arranged? Answers to such questions have taken many forms and have made for the most complicated aspect of the telecommunications business. There has never been a scarcity of coding schemes in the industry. Starting with the Morse code, to the Baudot code, and then the ASCII code, we have seen each providing for better transmission and higher quality. In this section, we will discuss the most popular and important three codes.

SONET

SONET is a standard for optical telecommunications transport. The SONET standard is expected to provide transport infrastructure for worldwide telecommunications at least for the next two or three decades. It defines a technology for carrying many signals of different capacities through a synchronous optical hierarchy. The standard specifies a byte-interleaved multiplexing scheme. The SONET standards govern not only rates, but also interface parameters, formats, multiplexing methods, and operations, administration, maintenance, and provisioning (OAM&P) for high-speed transmission. We most often hear of SONET rings in which fiber strands are strung around a metropolitan area in a ring configuration. The system is designed so that transmission can take place in either direction; should there be a fault at any one location, transmission will immediately take place in the opposite direction. That is, the system is self-healing.

Asynchronous Transfer Mode (ATM)

ATM is a high-performance switching and multiplexing technology that utilizes fixed-length packets to carry different types of traffic. Information is formatted into fixed-length cells consisting of 48 bytes (8 bits per byte) of payload and 5 bytes of cell header. The fixed cell size guarantees that time-critical information (e.g., voice or video) is not adversely affected by long data frames or packets. Of course, if the cells were longer the system would be more efficient, because the header would take up a smaller percentage of the total cell. Multiple streams of traffic can be multiplexed on each physical facility and can be managed so as to send the streams to many different destinations. This enables cost savings through a reduction in the number of interfaces and facilities required to construct a network.

Asymmetric Digital Subscriber Line (ADSL)

ADSL is, essentially, a modem that employs a sophisticated coding scheme. This coding scheme permits transmission over copper pairs at rates as high as 6 Mbps for distances of 9,000 to 12,000 feet. Speeds of this magnitude bring to mind television signals: a 6 Mbps channel can easily handle a television movie. ADSL succeeds because it takes advantage of the fact that most of its target applications (video-on-demand, home shopping, Internet access, etc.) function perfectly well with a relatively low upstream data rate; hence, the word asymmetric. ADSL is now used as an access technology for television businesses and for Internet access.

Another key technology changing the telecommunications landscape rapidly is switching. To understand the changes in switching technology, we shall look briefly at its evolution.

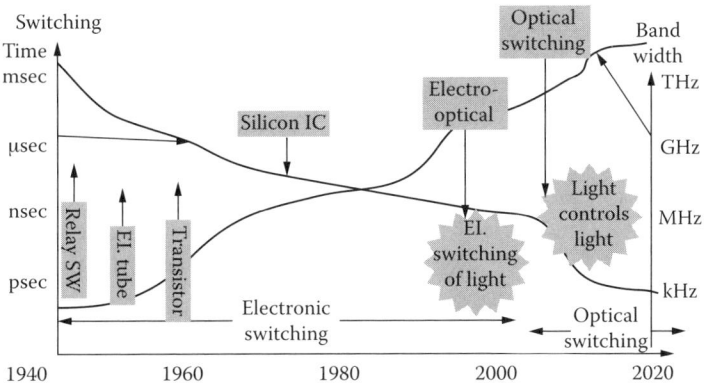

Figure 3.5   Evolution of switching technology.

Trends in Switching Technology

If there were only three or four telephones in a locale, it would make sense to connect each phone to all other phones and find a simple method of selecting the desired one. However, if there are three or four thousand phones in a locale, such a method is out of the question. Then it is appropriate to connect each other phone to some centrally located office and perform switching there.

International Engineering Consortium

The preceding quote explains the utility of switching technology in the telecommunications industry, as it is vital for an economically viable mass communication system.

The central office (CO) and switching technology have gone through a number of fundamental technological changes over the last 100 years. In the 1800s calls were connected manually at the central office. When a call came in, an operator would plug into a horizontal bar line and then shout out to the operator who handled the customer being called. The second operator would subsequently connect to the bar and finish setting up the call. Then came the step-by-step system that used a Strowger switch having an electromechanical operation which responded to the dial pulses of the rotary dial. The crossbar switching system of the early 1900s was also electromechanical in nature.

Later, the electromechanical control of the common control system was replaced with electronics. Major developments took place after 1947 when Bell Labs developed the first transistor. Existing networks were replaced with reed switches; however, only a part of the switch was electronic. In the next generation, the stored program operation of a digital computer was applied to the switch, though the network still remained a complex of reed switches.

The final generation was that of the digital switch, in which the talking path was no longer an electronically continuous circuit, rather, the speech being carried was digitized into a stream of 1's and 0's. Thus, this stage operates on a digital, not analog, domain, unlike the preceding stages.

The current trend has changed the whole switching operation in the telecommunications process and has driven the telecom revolution enormously. Going back to switching systems, there is a particular talking path from the calling party to the called party. This path was established at the beginning of a call and held for the duration. This is called *circuit switching*. Circuit switching is a type of communications in which a dedicated channel is established for the duration of a transmission. The most ubiquitous circuit-switching network is the telephone system, which links together wire segments to create a single unbroken line for each telephone call. We should note that circuit switching dominates the public switched telephone network (PSTN). A line is dedicated for a telephone call.

Circuit switching is not very efficient. In today's telecommunication, there is a different connection system called *packet switching*. Packet switching refers to a protocol in which messages are divided into packets before they are sent. The packets are sent individually and can follow different routes to the destination. Although packet switching is very efficient, it lacks in one aspect — sounding call. Circuit switching offers the best sounding call because all packets are transmitted in order. Delays in packet switching for voice cause voice quality to fall apart. However, as the telecom revolution goes forward and technology gets better, voice over packet-switched networks will get better.

The normal telephone service is based on circuit-switching technology; it is ideal when data must be transmitted quickly and must arrive in the same order in which it was sent, for instance, in the case of real-time data such as live audio and video. Packet switching technologies are useful for protocols such as wide area network (WAN), e-mail messages, and Web pages. This technology is suitable for the data communication needs of education, business, and governments.

In the long distance market, packet transport offers the promise of lower costs over traditional time division multiplexing (TDM) transport. This has resulted in an arbitrage play by upstart carriers, as they offer long-distance voice almost for free. Packet technology in the form of the Internet has also created major disruptions. As data traffic grew, forecasts of data surpassing voice abounded. With the continuing explosion of data traffic, the idea of a common transport gained momentum.

End-office switches were engineered for traditional voice-calling patterns. In fact, the steady increase in Internet traffic threatens to exceed these switches' capacity. The most cost-efficient method of providing a dedicated service, such as telephony, was to locate the necessary resources in the centrally controlled network. The X.25 standard for packet-switched networks was developed in the mid-1970s in CCITT, which is a standardization body of the PTOs. The effort was based on virtual circuits that kept track of each individual connection. The resulting behavior was similar to that of the traditional telephone networks, despite the introduction of packet switching. Reliability support and processing power were added to the network.

Simplicity and efficiency arguments led to the requirement that the operations of the subnets be based on identical technology, which was feasible owing to the monopolies in telecommunication. The complexity was hidden in the network, and generic services were offered to dumb terminals through predefined interfaces. These operating principles are in line with the operating principles of the centrally controlled telephone networks outlined earlier. Hence, X.25 technology — and later the ATM technology — represents a prolongation of the telecommunications community's perception, which has been heavily influenced by telephony.

One avenue of relief for this situation is the deployment of residential ADSL. The leading technology in this market is an always-on connection, but the traffic does not go through the local voice switch. Rather, the local Internet traffic is terminated in the local CO by a digital subscriber loop multiplexer or DSLAM. From the DSLAM the traffic is typically multiplexed into an ATM network and transported to an Internet Service Provider (ISP).

Today, this ATM network is separate from the voice network. Within the Internet infrastructure, as well as in the enterprise domain, the Internet Protocol (IP) is the dominant transport. Given that both ATM and IP have strong supporters, there is significant discussion and debate over what is the best approach, especially in replacing the legacy TDM network.

A New Era of Telecommunication

In 1962, MIT professor J.C.R. Licklider, who became the first head of the computer research program at the Defense Advanced Research Projects Agency (DARPA, later ARPA), put forward his galactic network concept. In the galactic network, much as in the modern Internet, users would be able to quickly access data and programs from any site.

In 1964, while researching secure communication via packet switching with colleagues W. D. Davies and Paul Baran, Leonard Kleinrock, also of MIT, published the first book on the subject, *On Communication Networks* based on a 1961 paper about packet-switching theory. Their research confirmed the theoretical feasibility of communication using information packets rather than circuits, an important step toward computer networking.

A key step was to make computers talk. In 1965 researchers connected the TX-2 computer in Massachusetts to the Q-32 in California with a low-speed dial-up telephone line, creating the first wide area computer network. Although the experiment was considered a success, the communication itself was slow, expensive, and inefficient. It did confirm the theory that computers could work well together, both to run programs and retrieve data. Importantly, it also proved, as Kleinrock had predicted, that packet switching would be the most promising solution to computer networking. The circuit-switched telephone system was totally inadequate for the job for reasons of speed and scalability.

The Network Is Born

The year 1966 marked the birth of the ARPANet, a government-funded project aimed at establishing a network that, among other objectives, would test the packet-switching theory. Systems were added to the ARPANet for the next few years at the rate of 1 per month and by the end of 1971, 19 nodes were connected to the ARPANet.

In 1970 ARPA's Network Working Group (NWG) completed the preliminary ARPANet host-to-host protocol, Network Control Protocol (NCP), which systematized the way computers talked to one another, this meant that network users could finally begin to develop network applications. A year later, NWG finished the Telnet Protocol and began work on the File Transfer Protocol, both of which are still in use today.

In March 1972, Ray Tomlinson of BBN, a company closely involved with the evolution of ARPANet, was motivated by the need of the ARPANet developers for an easy collaboration mechanism and developed a basic e-mail send-and-read software. ARPA's Laurence Roberts expanded the capabilities of e-mail by writing the first e-mail utility program to list, selectively read, file, forward, and respond to messages. From there, e-mail took off as the largest network application for over a decade.

Comparing the Telecom and Data Network Models

Data communications is the means by which our evolving culture has implemented an exchange of intelligence in our society. Exchange of

Table 3.1   Characteristics: Voice Network versus Data Network

Voice Network	Data Network
Mainframe-based management (AIN: advanced intelligent networks)	Mainframe management in the core, with client/server (C/S)-based management at the edge
Monopoly-based solutions (one provider per country)	Open solutions for the subscriber (many providers per region offering various levels of services)
Intelligence in the backbone	Intelligence throughout the network
Highly scaleable to hundreds of thousands or even millions of users; highly reliable and stable	Highly scaleable; mainframe scalability to millions of users; C/S scalability to tens of thousands of users
Circuit switching	Packet switching
Poor support for data transfer over the voice network (e.g., nonguaranteed speeds of 56 Kbps V.90 modem communications)	Excellent support for voice and video over data networks (e.g., H.323 support with QoS)

intelligence has always been and continues to be the dominant force that dictates the way we live and do business. Let us examine why these two models are different.

Model

The telecom model is basically characterized by the fact that the communicating parties have to go through the mediation of the network that controls the communication service. This is not linked to technology (for instance, the availability of simpler user devices), but to the fact that the network's basic raison d'être is to offer this communication service.

A data network, though fundamental for the proper operation of the closely interconnected computing devices of today, is not built with the objective of supporting a specific application. The network's duty is to transport data for multiple applications hosted by the computing devices involved in communication. Those applications know and control the communication service end to end. In terms of the Open System Interconnection (OSI) reference model, the application

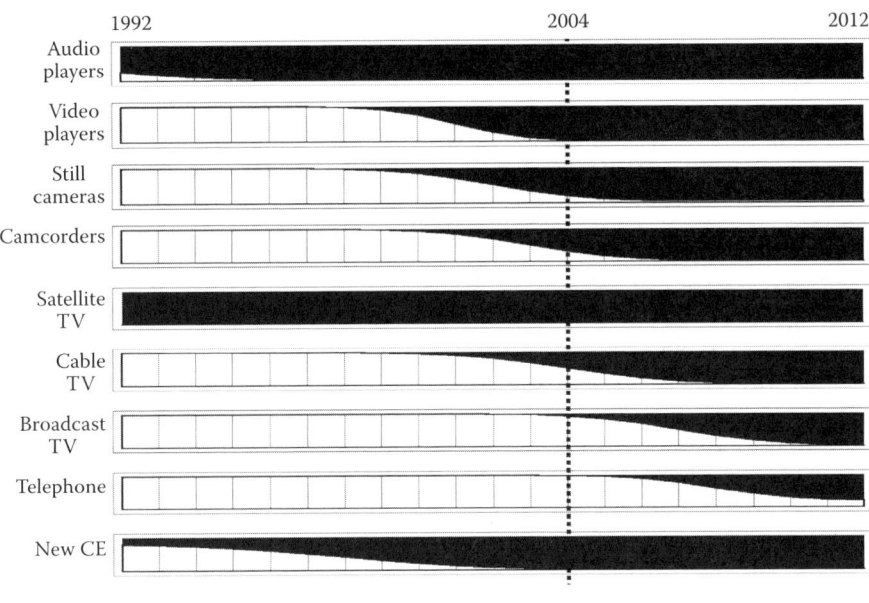

Figure 3.6   A digitized world.

layer is hosted by the communicating devices that directly communicate between each other (hence the term *two party*), and the network seldom supports beyond the transport layer.

Protocols

In a telecom network, the devices that route the user's traffic within it interpret the semantics of the communication service; their role goes beyond basic routing to managing user service requests and coordinating with peer devices for a proper completion of those requests. The user-to-network dialogue for service requests and the dialogue between network devices for proper service completion are defined by protocols. These protocols are specified and agreed upon by subnetwork operators worldwide and are not known to end-user devices — except for the role of such protocols in setting up the user-to-network dialogue.

In a data network, the devices used to route user traffic essentially address transport issues and are limited to the role of routing a given piece of data — generally, a packet — when presented, to its destination or the network device closest to it. Once this operation is performed no state information is kept within the device. Any user-network or network-level protocol used for a given application is specific to it, and is not an inherent part of the network as in the telecom model.

Quality of Service

In a telecom network, as the service is managed by a network whose operator derives revenue from its provision, service denial is preferred to bad service. Good service quality is guaranteed by proper network resource reservation all along the path linking the communicating parties. Quality of service — good service and service denial reduction — is closely linked to proper network dimensioning both at the access and core network (within an operator network and at interoperator network boundaries) levels. The cost of quality of service is not only connected to the actual transport resources reserved for a given communication but also, and more important, to the involvement, i.e., dialogue and state maintenance for the call duration of network devices.

In a data network, user traffic is delivered through specific interconnection points. Such points are generally associated with a service level agreement (SLA) that determines the general properties of the transport service that the network can support in terms of service availability, transport reliability, average or peak data rate, delay classes, etc.

Accessibility and Universal Reach

Access to the worldwide telecom network is provided by every subnetwork operator (or public network operator) to his or her subscribers. Operators generally charge a basic fee (subscription) that barely covers the costs of access provision, the bulk of their revenues being generated from the service usage triggered by a universal and ubiquitous access. Access universality is guaranteed by the interconnection agreements that link operators; the benefits of augmentation of subscriber base by one operator are automatically shared among other operators by augmentation of the global number of subscribers that can be reached through the telecom network. Universal access, of course, implies the existence of an addressing scheme that is consistent and universally acceptable; it also implies mutual operator obligation for proper completion of calls with good quality within their network (whether they are used only to transit a call or to complete it to final destination).

Data network deployment follows a pattern that is distinct from that of telecom networks. As revenue is not generated by selling end-to-end services, but basically by offering a general-purpose data transport service, it is natural that data networks, whether private or public, addressed corporate or academia users almost exclusively. The internetworking of data networks through the IP protocol for the support of Internet applications was initially limited to the aforementioned users. When the need arose to offer the general public access to Internet applications, with the introduction of personal computers sufficiently powerful to host them, the easiest solution was the use of modems and connections through the telecom network because of its ubiquitous access, especially in developed countries.

Interconnection agreements between data operators being independent of applications, universality of reach is provided on per-application basis in an ad hoc manner. Of course, each network provides a transport address (for instance, an IP address) to all of its connected users; such addresses, however, may have only local significance or may not be permanent. Therefore, many applications, such as Web browsing or e-mail, use symbolic addresses that are translated with the cooperation of decentralized network servers that translate the symbolic name into a valid transport address for the destination. Therefore, a network interconnection agreement per se is not sufficient to ensure access universality; each user and network operator has to determine (for each application) the name of the translation server that has to be addressed for a proper completion of a communication.

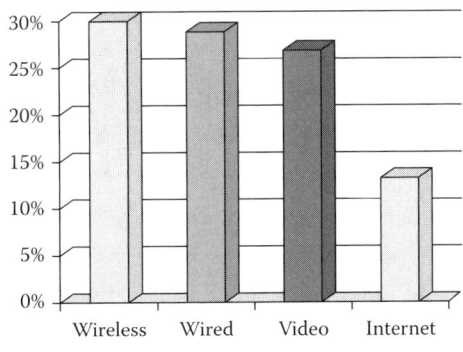

Figure 3.7   Consumer telecommunications spending.

Charging

As a data network service is basically a transport media, it is natural to charge the volume of the transported data or even a flat rate that generally gives the right to send or receive a given amount of data. A major and widespread simplistic misconception nowadays consists of opposing the per-minute charging model of the telecom network to the volume (or flat-rate) charging model of data networks as being related to their respective transport technologies (circuits versus packets), the latter mode being considered as more effective especially for similar services (voice transport). First, as explained earlier, voice telephony is not only voice transport. It is above all a service with many other attributes (provision of an address, permanent network access, network operator responsibility for proper service completion, etc.). Its per-minute charging model is more linked to its nature as a high-level application service rather than to its support by a circuit-switched technology. Second, on the other hand, data network charging is based on volume or flat rate because of the nature of the service provided, that is, data transport. The network-added value here is basically that of transporting a given volume of data from one point to another. It is therefore natural that the charging metric be related to that volume irrespective of the technology used.

From the strict economic point of view, the need to clearly understand these two charging models and their impact on both operator and service provider revenues is important.

The merger of the two communication models places the revenue value chain at the heart of the debate as a prerequisite for the successful, widespread merging of IP-based transport and user applications in the current telecom environment.

Chapter 4

The Internet Takes Off

> The new information technology, Internet and e-mail, have
> practically eliminated the physical costs of communications.
> There is now a low-cost technology, finally, that for the first
> time in human history allows people to really maintain rich
> connections with much larger numbers of people.

The new ARPANet network technology was introduced to the public
in 1972 at the International Computer Communication Conference
(ICCC). From there, the Internet rapidly evolved out of the idea that
multiple independent networks of arbitrary design would join the
ARPANet, the first packet-switched network. It could then accommodate other networks, such as packet satellite, ground-based packet
radio, and other networks.

Open-architecture networking, the fundamental idea that underlies
the structure of the modern Internet, was first proposed by Bob Kahn,
at ARPA. Kahn defined *Internetting*, as it was called at the time, by
four critical ground rules:

- There would be no global control at the operations level.
- Each distinct network would have to stand on its own, and no
  internal changes could be required of any such network to
  connect it to the Internet.
- Communications would be on a best-effort basis. If a packet
  did not make it to the final destination, it would be retransmitted
  from the source.

■ Black boxes, later called *gateways* and *routers*, would be used to connect the networks. These black boxes would not retain any information about the individual flows of packets passing through them.

Other key issues included developing algorithms to prevent lost packets from permanently disabling communications, providing for host-to-host pipelining so that multiple packets could be en route from source to destination as allowed by participating hosts and intermediate networks, and a network-to-network protocol. Kahn and Vint Cerf (a Stanford researcher who headed the first International Networking Working Group), presented their first paper on the new internetworking protocol, TCP, in 1973. Three years later they demonstrated Internetting in public demonstrations in which they interconnected a packet radio network, SATNET, with the ARPANet.

By 1984 over 1000 hosts were connected to the nascent Internet. To make it easy for people to use the network, each host machine was assigned a name so that it would not be necessary to have to remember each host's numeric address. When the network consisted of only a few hosts, each host maintained a single table of all the hosts and their associated names and addresses. With an increasing number of independently managed networks (e.g., LANs), a single table of hosts was no longer feasible. To maintain network scalability without introducing the administrative nightmare of managing such a large hosts table, Paul Mockapetris of University of Southern California Information Sciences Institute (USC/ISI) developed the Domain Name System (DNS). The DNS permitted a scalable distributed mechanism for resolving hierarchical host names into an Internet address. This simplified administration, because by 1987 the Internet had grown to 10,000 sites. By 1992, this number would grow to over a million, and by 2005 touched a billion. Early networks were built for and largely restricted to closed communities of scholars.

The Birth of the Commercial Internet

Starting in the early 1980s and continuing to the present, the Internet grew beyond its primarily research roots to include both a broad user community and increased commercial activity. Originally, commercial efforts mainly consisted of vendors providing the basic networking products and service providers offering connectivity and basic Internet services. The Internet has now become a commodity service, and much of the latest attention has been focused on the use of this global information infrastructure for support of other commercial services. This has been tremendously accelerated by the widespread and rapid integration of the World Wide Web and browser technology, allowing users easy access to information linked across the globe. Products that facilitate the distribution of that information and many of the latest developments in technology have been aimed at providing increasingly sophisticated information services on top of the basic Internet data communications.

The Internet has its roots in the electric telegraph; these widespread means of communications mirror each other at different stages of their development in significant ways. The first Internet, a natural outgrowth of this mode of personal communication, was built on the many advances made by the telephone industry (i.e., the transistor, microwave radio relay, cable video transmission, teletypewriter networks). Conceived to allow remote logins and data retrieval, today's Internet is evolving to support services requiring even higher bandwidth, such as streaming audio and video, video telephones, and teleconferencing. Modern networking combined with powerful yet inexpensive laptop computers, two-way pagers, PDAs (personal digital assistants), and cellular phones is making possible a new paradigm of J.C.R. Licklider's galactic network vision with high-speed, portable, and mobile communications. The continuing development of the Internet even as these words are being written is proof that the Internet's evolution is not yet complete.

Related Networks

In 1980–1981, two other networking projects, BITNET and CSNET, were initiated. BITNET adopted the IBM RSCS protocol suite and featured direct leased line connections between participating sites. Most of the original BITNET connections linked IBM mainframes in university data centers. This rapidly changed as protocol implementations became available for other machines. From the beginning, BITNET has been multidisciplinary in nature with users in all academic areas. It has also provided a number of unique services to its users (e.g., LISTSERV). Today, BITNET and its parallel networks in other parts of the world (e.g., EARN in Europe) have several thousand participating sites. In recent years, BITNET has established a backbone that uses the TCP/IP protocols with RSCS-based applications running above TCP. CSNET was initially funded by the National Science Foundation (NSF) to provide networking for university, industry, and government computer science research groups. CSNET used the Phonenet MMDF protocol for telephone-based e-mail relaying and, in addition, pioneered the first use of TCP/IP over X.25 using commercial public data networks. The CSNET name server provided an early example of a white pages directory service, and this software is still in use at numerous sites. At its peak, CSNET had approximately 200 participating sites and international connections to approximately 15 countries. In 1987, BITNET and CSNET merged to form the Corporation for Research and Educational Networking (CREN). In the fall of 1991, CSNET service was discontinued after having fulfilled its important early role in the provision of an academic networking service. A key feature of CREN is that its operational costs are fully met through dues paid by its member organizations.

Internet Evolution

The Internet has functioned through collaboration among cooperating parties. Certain key functions have been critical for its operation, not the least of which is the specification of the protocols by which the components of the system operate. These were originally developed in the DARPA research program mentioned earlier, but in the last five or six years, this work has been undertaken on a wider basis with support from government agencies in many countries, industry, and the academic community. The Internet Activities Board (IAB) was created in 1983 to guide the evolution of the TCP/IP protocol suite and to provide research advice to the Internet community. During the course of its existence, the IAB has reorganized several times. It now has two primary components, the Internet Engineering Task Force and the Internet Research Task Force. The former has primary responsibility for further evolution of the TCP/IP protocol suite, its standardization with the concurrence of the IAB, and the integration of other protocols into Internet operation (e.g., the Open Systems Interconnection protocols). The Internet Research Task Force continues to organize and explore advanced concepts in networking under the guidance of the IAB and with support from various government agencies.

The Technology behind the Internet

Depending on the access technology and capacity of the network, Internet connectivity can be both extremely slow and, in many parts of the world, quite expensive. In this section we try to explore various Internet connectivity technologies that can be used to connect individuals, businesses, or other groups.

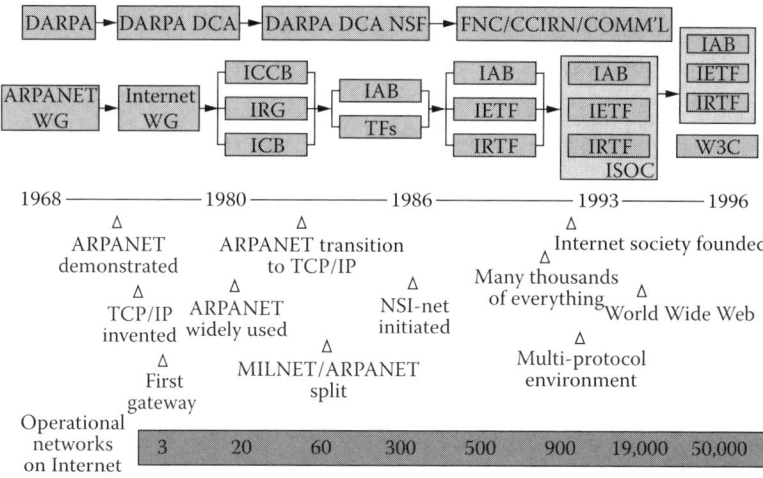

Figure 4.1   Internet timeline.

This section is necessarily incomplete, because the range of possible technologies that enable Internet connectivity is extremely broad. Because this field is the center of attention, not only for the global technology community but also for establishments, businesses, and nonprofit organizations, it is very dynamic and continuously evolving. Further, with some technologies it is not even possible to arrive at a conclusion, because new developments are always around the corner. This section, however, provides adequate information that can be used to determine which technologies might be appropriate for a particular scenario, and hence prepare the reader for further investigation.

Providing Internet access to a geographically dispersed set of locations requires two complementary network infrastructure components: a backhaul connection to the region, and a distribution mechanism to make the connection available to individual locations.

Backhaul Technologies

Backhaul technologies provide the main channel to the global Internet.

Backhaul connections can range from a dial-up Internet connection to a broadband DSL or cable modem to a high-end satellite solution. The right choice for a particular application will depend on bandwidth requirements, the budget, and available network capacity (Internet access from ISPs or telecoms) in the region. For example, a project providing connectivity to a set of kiosks that are located within a city

Table 4.1   Access Method versus Data Speed

Access Method	Data Speed (bps)
56K modem	56K
ISDN	128K
DSL	1.5M
Cable modem	1.5M
T1 line	1.544M
Microwave wireless	6M
T3 line	43M

will most likely be able to use a wired broadband connection to the city and then implement other wireless access technologies to distribute bandwidth to the kiosks.

On the other hand, a project delivering access to rural villages that are nowhere near a cable or DSL-equipped distribution point will need to make arrangements for a backhaul connection specifically for the project. In such a case, microwave or satellite connections (while expensive) are often the only choice for backhaul connectivity to the Internet.

Table 4.2 summarizes the various strengths and weaknesses of the major technologies that can be used for backhaul connections to the Internet.

Distribution Technologies

Once Internet connectivity is available in the region where a subscriber is situated, the bandwidth must be distributed to the final locations where it will be used. Distribution technologies take the bandwidth from the backhaul connection and make it available to users at one or more locations.

The choice of distribution technology will depend on the location of the endpoints of the network (including their proximity to each other), the topography of the geographic area, and the local communications infrastructure. In many cases (especially underdeveloped regions), the local communications infrastructure is weak or nonexistent; for these situations, the project will often need to deploy its own network infrastructure.

Distribution networks can be built from a wide variety of technologies ranging from wired local area networks to wireless networks to

Table 4.2 Backhaul Technology

Technology	Strengths	Weaknesses	Costs
Satellite	Can be installed virtually anywhere; high bandwidth available	High costs; difficult installation	High setup cost; high monthly cost depending on bandwidth
Microwave links	High data rates; covers distances up to 30 km; easy to provision new service	Requires additional backhaul to feed microwave network	Higher setup cost compared to satellite; high monthly cost depending on bandwidth but low compared to satellite
Wired broadband (cable and DSL)	Broadly available in urban areas; relatively low-cost; no need for new infrastructure; high bandwidth	Not available in many areas	Depends on location; low monthly cost
Dial-up	Available almost anywhere there is a phone line; no need for new infrastructure	Low speeds; not stable	Low setup cost; high monthly cost where calls are metered per minute

caching systems. For example, if a project simply needs to provide access to a set of workstations located in the same building, a simple wired Ethernet network will suffice for distribution. Distributing bandwidth to a set of villages that are several miles apart, on the other hand, will most likely require a wireless or caching solution that eliminates the need for physical wiring between locations.

Some technologies, such as wireless local loop, take advantage of existing backhaul infrastructures and are available from existing telecommunications providers. When available, wireless local loop can be used to provide distribution of bandwidth across a wide geographic area while offloading the infrastructure and network requirements to the provider.

Table 4.3 summarizes the various strengths and weaknesses of the major technologies that can be used for backhaul connections to the Internet.

Table 4.3 Distribution Technology

Technology	Strengths	Weaknesses	Costs
Wired LAN	Easy to install; low-cost; high data rates; well-understood technology	Requires near proximity to backhaul; cannot cross most distances	Very low – requires some networking hardware
Wired WAN	Reasonably high data rates; can cover long distances	Requires local communications infrastructure	Varies by distance and location; generally relatively high
802.11-Based wireless	Ideal for distribution within a small geographic area (such as a village); relatively low-cost; no communications infrastructure required; can use special hardware to extend range; mature technology	Limited range (200 m) for standard hardware; crossing long distances requires special hardware at higher cost	Basic hardware costs low ($50 for a wireless card, $50 for a wireless base station); high-gain antennas more expensive ($200–1000)
802.16-Based wireless	Long range (30+ km); high data rates (up to 70 Mbps); no need for local communications infrastructure; ideal for distribution to moderately distant locations	Unproven technology	Unknown – certified products not yet available

Mesh networks	Extends the range of wireless technologies; small scale-up increment; may be used to create a more robust distribution network	Relatively immature technology; must be combined with other technologies	In the range of $500 per mesh access point; mesh networks also require client hardware (generally 802.11 based)
Wireless local loop	Infrastructure built and maintained by telco; deployment quick and easy; reasonable data rates	Costs do not scale well; limited bandwidth; technology is somewhat immature	Low
Caching technologies (such as first mile solutions)	Virtually no infrastructure required; range only limited by vehicle range	No direct Internet connection	FMS case study in Cambodia set up villages for about $600/village

We will cover dial-up Internet connectivity using a modem in this section, and the remaining technologies will be covered later in this chapter. It has long been thought that the theoretical limit on modem speed over an ordinary phone line was 33.6 kbps. 56K modems achieve their speed by avoiding a conversion from digital to analog lines in the connection between user and service provider. Ordinary connections begin over an analog line, are converted to digital by the phone company and are converted back to analog in the final segment before arriving at the service provider. 56K connections begin analog, are converted to digital and are not converted back to analog at the service provider. This requires the service provider to have a direct digital connection and therefore avoids one conversion of the signal. By avoiding this second conversion, speeds of up to 56 kbps and higher are possible. Therefore, modem users need to know that they can only achieve 56K if their service provider supports it.

When the 56K fever first affected the industry in the fall of 1997, there were three camps — Lucent Technologies, Rockwell International Corp., and U.S. Robotics — each proposing its own proprietary 56K modem specification. Lucent and Rockwell joined forces to design the K56flex specification. But U.S. Robotics already had a considerable head start and was able to ship its x2 products several months ahead of the competition.

Partly as a result of a merger with 3Com (a supporter of K56flex) earlier in 1998, U.S. Robotics agreed to work with Lucent and Rockwell on a unified, worldwide 56K modem standard under the ITU. In 1998, both the K56flex and x2 camps abandoned their existing 56K specifications and switched over to this new, unified standard — V.90.

Unprecedented Internet Growth

Since the 1990s when the Internet began to impact the way people communicate, there has been an explosion of unprecedented scope in the usage of the Internet. Its impact on how the world communicates, gathers information, and does business cannot be overestimated.

The global Internet continues to evolve. Consumer and business Internet subscriptions are steadily increasing, and broadband demand is experiencing explosive growth. The most promising growth opportunities are the new markets: new regions, new technologies, new business models, and new partnerships.

Contrary to popular belief, Internet access affects every member of society, including those seemingly most remote from the technology.

In the developing countries, the Internet has quickly emerged as one of the most useful means of communication, definitely far more useful than the telephone. In practice, the Internet adds more services, many of which are especially appreciated in areas of extreme poverty where people are isolated from health, government, education, and other facilities, making access to information crucial. In practical and economic terms, a document sent by e-mail costs significantly less than a fax, whereas international telephone calls made via the Internet enable expatriates to be contacted at very low cost. Lastly, cable, wireless, and satellite infrastructures can bring isolated areas, such as Africa, out of the wilderness and connect them to the world's high-speed communication backbones.

The three key areas of growth have been in the use of the Internet by businesses and individuals, the growth of E-commerce and online business transactions and communication, and the use of search engines that collate the billions of Web pages that make up this vast resource. The Internet has great transformative power over the way companies do business, the way individuals communicate, and the manner in which people spend their recreational time. It is the applications that will drive Internet evolution, creating demand for ubiquitous bandwidth.

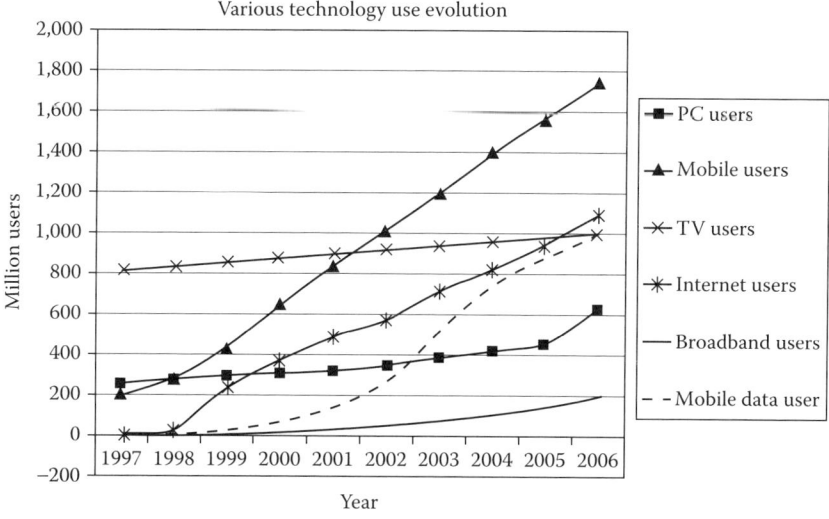

Figure 4.2   Number of users for various technologies.

There has been a huge, worldwide explosion in the use of the Internet. It is already changing the way we work, shop, bank, and also the way we live. The PC has become commonplace in many homes and almost all offices. Access to the Internet and its services has already become widespread. The use of the Internet by many businesses, organizations, and the general public has led to a rapid increase in E-commerce. Clearly, all these developments are going to change the way we work and live.

The Impact of the Internet

The Internet has the potential to spread access to knowledge, information, and markets to people who have traditionally been excluded from these crucial aspects of development.

The Internet has increased the power of connectivity because of the wealth of resources it has made available to those who are connected and have the skills to tap this abundant resource. It can enhance the capability of those with access in the fields of education and health, as well as providing new opportunities for economic activity and democratic participation.

The Internet has indeed changed everything. Beyond the hype and the headlines exists the real work of building the new business structures and technologies that can ensure success in a rapidly changing business world. Today, companies are evaluated as much on their ability to adapt to the Internet as they are on their prior performance. This pressure has been created by a worldwide realization that the Internet is fundamental to the way the world will work. Thus, companies that are not using the Internet intelligently are not working intelligently at all.

Net paradigms of work are emerging in business industry and society. This may eventually lead to a transformation of individual lifestyle itself.

Society and Individuals

If information explosion is the most prominent phenomenon of the present era, the Internet is, and will remain, its greatest enabler. This explosion is triggering a global knowledge revolution. The role of the Internet in driving this revolution will be tremendous. Its impact will be far greater than that of all information dissemination technologies in use since the Industrial Revolution. And that is because of the Internet's unique power to deliver any information to anyone, anywhere smartly, instantaneously, and inexpensively. A significant impact of this power will be on the interconnectivity of people, regardless of geography and time zone. In fact, this impact is already discernible.

The rise of the Internet has dramatically changed the global village. The Internet has created a network over which far more than voice, i.e., text, video, and other data in intelligent form can be exchanged. The global village has now gone online, making the villagers active participants and not just spectators as was the case in the past. The Internet has reduced the delay between stimulus and response, enabling ordinary people to share feelings, ideas, and reactions to events as they happen or within minutes.

The Internet is a provider of unlimited knowledge. We increasingly use it as a universal reference library. It has transformed education, research, and our daily needs for information. Type any keyword into a search engine and the chances are you will find a host of sites with information on that topic. With the provided links, you can find further details.

In cyberspace we meet people from diverse backgrounds and in some cases create virtual relationships. One major impact that the Internet has had on society is in transforming the field of personal relationships. The Web has taken the old models of personal advertisements and dating agencies and catapulted them into new dimensions. The new breed of online dating sites allow people to "meet" at a distance, chat in real-time or by e-mail, exchange photographs, and even "talk" to each other with voicemail. Distance is no longer a barrier — many international romances have blossomed on the Internet. Chat rooms are another extremely popular way for meeting people from all over the world online. People can discuss whatever they wish and join or leave whenever they choose to.

Today's connected world makes access to expertise in many areas a reality even for poor people in distant locations. Remote intervention by the best doctors, consultants, and teachers based in other parts of the globe has made the dreams of telemedicine and distance learning (or telelearning) a reality. The Internet and multimedia have the potential to transform healthcare, education, and business by removing the need for doctor and patient, teacher and student, and executive and office to be in the same place.

It seems strange now that we managed without e-mail for so long, using telephones and the mail. The impact of e-mail has been astounding, transforming information delivery and making it possible to keep everybody informed about everything all the time. Also, e-mail can be easily termed the most important method for communicating and developing relationships since the telephone. In a sense, it takes us full circle. Our grandparents used to keep in touch by letter before the telephone age sent correspondence with text into what seemed like terminal decline. Now it is back — but in a different form. The e-mail medium is more considered and deliberate than a telephone call. You have time and space to shape your thoughts into words. But it is much faster than writing a letter, and you can get a reply in minutes if the other party is online too. You can share e-mails among groups, use them to exchange pictures, sound, or video files, or links to favorite Web sites. Effectively, e-mail is much more than electronic mail sent over the Internet. It creates a separate psychological environment in which pairs (or groups) of people can interact agreeably, providing a context and boundary in which this interaction can unfold.

Enterprises and Businesses

The Internet is turning business upside down and inside out. It is fundamentally changing the way companies operate, whether in the high-tech sector or cement and steel, whether operating from Manhattan or from a hamlet in Africa, and whether the company is a corporation or small and medium enterprise (SME). Although E-commerce does have a lot of promise, this goes far beyond buying and selling over the Internet, and profoundly impacts the processes and culture of an enterprise.

The Internet has created a challenge for every area of every company. The challenge is not simply to change one aspect of how a business operates, it is to change every aspect. All of this is occurring against a backdrop of competitors and start-ups hoping they can use the Internet to be better than their competitors. Keeping pace with the Internet thus becomes critical; it is easy to fall behind, especially if you develop your plans in a vacuum, failing to watch and learn from the best-of-breed Internet successes.

Today, more and more companies are using the Internet to make direct connections with their customers or are using it to strengthen relations with their trading partners, and vendors are using the Internet's reach and ubiquity to request or provide real-time information and buy or sell stocks of goods or services directly or by auctions.

Entirely new companies and business models are emerging in industries ranging from commodities to logistics to bring together buyers and sellers in superefficient new electronic marketplaces. The Internet is helping companies lower costs dramatically across their supply and demand chains, take their customer service into a different league, enter new markets, create additional revenue streams, and redefine their business relationships.

As the age of the Internet unfolds, many businesspeople and consultants see the key to success as being as much about the right plan, the right business model, and the right corporate structure as it is about Web pages and fancy Internet infrastructure. Every company must learn to change and implement changes so that it can take advantage of the new ways in which the Internet allows companies to market products, obtain supplies, provide customer service, and interact with business partners. Integration is the true key. The more you can use the Internet to tie together your corporate infrastructure, goals, and technology, the more successful you can be. There are many positives and opportunities, but three key proven facts about the Internet's effect on commercial activity are the following:

- First, it shifts power from sellers to buyers by reducing the cost of switching suppliers (the next vendor is only a mouse click away) and freely distributing a huge amount of price and product information.
- Second, the Internet reduces transactions costs and thus stimulates economic activity. A banking transaction via the Internet costs 1 cent, compared to 27 cents at an automatic teller machine (ATM) or 52 cents over the telephone. Processing an airline ticket on the Internet costs $1, compared to $8 through a travel agent.
- Third, the speed, range and accessibility of information on the Internet and the low cost of distributing and capturing it create new commercial possibilities.

The Internet will also bring about revolutionary enhancements in productivity. The magnitudes of productivity increase will indicate this transition to the Internet economy. During 1990–1995 there was a 1.6 percent improvement in productivity, in terms of GNP. During 1995–1998, this was 2.6 percent. In 2000–2003 it was 3 percent and is likely to be 5 percent by 2008. In the United States, the network effect has improved productivity by 50 percent, whereas in Japan it is

20 percent. An Internet solutions company such as Cisco has attained a year-on-year growth of 58 percent. Its market capitalization is $442 billion and the revenue per employee is $700,000, which is unmatched in the industry; the next best is $250,000.

According to research conducted on about 250 top U.S. companies by Momentum Group last year to understand the impact of being networked, this growth in use of the Internet, in turn, is driving the creation of new applications and new ways of doing business. Several new technologies will mushroom to harness the full potential of the Internet. Hardware will get cheaper, global use of the Internet will confer a competitive advantage, employee empowerment will increase, everything will be customer driven, and change management will be reckoned as a distinct capability.

The resulting demand for reliable, high-performance bandwidth has enabled service providers to grow revenue, even though retail prices for bandwidth, on a per-megabit basis, have dropped drastically during the past five years. Indeed, service providers are faced with a formidable challenge: grow profits in an environment in which technology improvements guarantee that bandwidth prices will continue to decline.

The high-tech bubble was inflated by myths of astronomical Internet traffic growth rates. Although these myths were false, Internet traffic has been increasing very rapidly, almost doubling each year since 1997. Moreover, it continues to keep growing close to this rate. This rapid growth reflects a poorly understood combination of many feedback loops operating on different time scales.

Chapter 5

The Broader the Better

Broadband refers to high-speed always-on connections to the Internet that support the delivery of innovative content and services. Compared to traditional narrowband connections, broadband access is immediate and large volumes of data can be almost instantly transmitted, reducing waiting time and improving efficiency for users.

Downloading large software files or video or audio files on narrowband can be a time-consuming and frustrating exercise. By using a broadband high-speed Internet connection, with data transmission rates many times faster than a 56K modem, users can view video or download software and other data-rich files in a matter of seconds. In addition to offering speed, broadband access provides a continuous always-on connection (no need to dial up) and a two-way capability, that is, the ability to both receive (download) and transmit (upload) data at high speeds. Broadband access, along with the content and services it might enable, has the potential to transform the Internet — both what it offers and how it is used.

Broadband plays a major role in modernizing economies and societies. As an enabling technology, it is at the core of the diffusion of the information society and of the development of information and communication technologies (ICTs). These technologies in turn are key drivers of productivity and growth.

Broadband enables the delivery of new, advanced content. It promotes the development of new services and improved delivery of those that already exist. It allows the reorganization of working and production processes. All these developments bring significant benefits to businesses, administrators, and consumers.

The benefits of broadband are widely recognized. Many developed nations are already exploiting these benefits as they experience significant increases in deployment and take-up, which is largely market driven. There are, nevertheless, obstacles to more rapid progress.

Fueled by recent steep technology price drops, businesses of all sizes accelerated the pace of PC acquisition to make their employees more efficient. The need for multiple workstation connectivity, combined with the advent and proliferation of broadband Internet access, brought about dramatic performance gains for businesses around the world: no more waiting for the dial-in process, faster navigation, and quicker downloads.

Although residential broadband adoption is rising sharply, the roll-out of broadband access technologies to residences is still in the early stages in most countries. With broadband access becoming available to home users, the way people work and play is changing. What they are finding is that not only does broadband access result in fast Web surfing owing to higher connection speeds, but that there are also several other benefits.

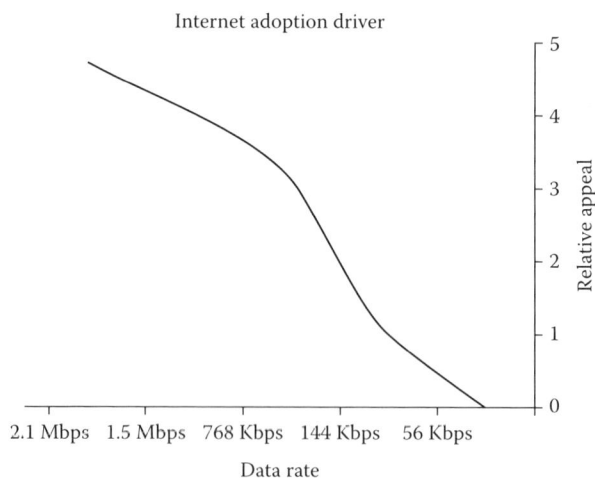

Figure 5.1   Data rate as Internet adoption driver.

The Need for Broadband

> The residential broadband market will grow to reach $80 billion by 2007. By then, there will be almost 300 million businesses and homes with broadband around the world, outnumbering those businesses and homes with dial-up access.
>
> **ARC Research Group**

Further, the same research indicates that over 30 million of the broadband connections will be to businesses. Most of the industry observers and analysts see enormous scope for future growth in the broadband market, expecting in excess of 400 million broadband connections by the year 2010. These numbers definitely look too optimistic to achieve. However, they seem practical on a relook in light of the fact that most developed nations have household broadband penetration rates of no more than 20 percent, whereas for developing nations they are not even worth writing home about.

Another interesting observation made by researchers is that though both enterprise as well as residential broadband markets will grow substantially, residential broadband connections will emphatically outnumber enterprise connections in future. The enterprise market with fewer connections will be larger in terms of revenue and bandwidth usage. North America will lead the broadband business market, followed by Western Europe and Asia Pacific.

Broadband affords end users high-speed, always-on access to the Internet while affording service providers the ability to offer value added services to increase revenues. Compared to traditional narrowband connections, broadband changes the overall presentation of the Internet, from slow and often user-unfriendly text format to a fast, colorful system combining video, animations, and sound. Connections are immediate and large volumes of data, notably video and graphical content, can be almost instantly transmitted.

Broadband opens the way to the creation of new markets through the development of increasingly interactive applications and new high-quality services. Beyond the emergence of new multimedia applications, a wide range of services is expected to grow in parallel with the take-up of broadband, delivering new economic and social benefits.

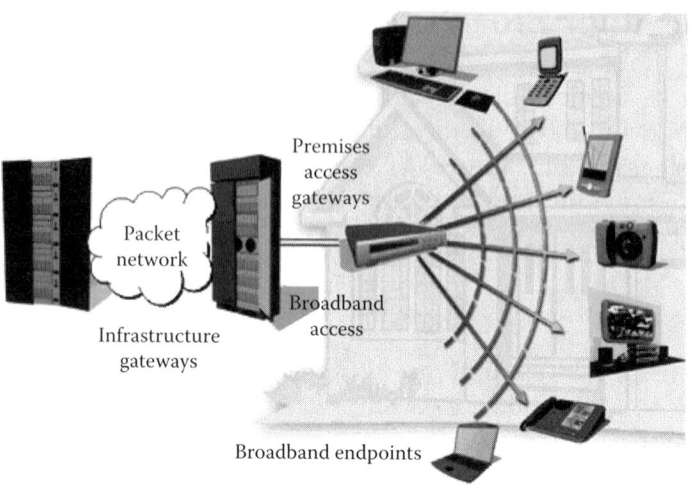

Figure 5.2 Basic broadband architecture.

Why Broadband?

According to various surveys and polls conducted across the globe, key reasons cited by users for switching to broadband were the following:

- Existing connection too slow
- Faster file downloads
- Job-related tasks
- Always-on connectivity
- Simultaneous phone/Net usage
- Higher-quality connection
- Easier access to entertainment
- Responded to promotion
- Affordable price

Broadband Applications

The high speed and high volume that broadband offers could also be used for a bundled service offering, for example, cable television, video-on-demand (VoD), voice, data, and other services over a single line. It is possible that many of the applications that will best exploit the technological capabilities of broadband, while also capturing the imagination of consumers, have yet to be developed.

Enterprises and Businesses

Broadband provides a myriad of ways to improve the efficiency and productivity of enterprises and businesses while achieving substantial cost savings at the same time. Some of the ways to take advantage of the benefits of broadband are the following:

Telecommuting and distributed enterprise: Communicate with the main office over a secure broadband link. Your office data is at your fingertips; broadband makes the corporate network available from remote locations. Using broadband to implement a virtual private network, an employee can work from home or a remote location and have the same access and level of response as if he or she were seated in the office.

Voice-over-IP (VoIP) or Internet telephony: VoIP is the latest cost-effective means of communication, made more feasible by broadband. It allows you to make telephone calls and send faxes over data networks such as the Internet and intranets. Although it currently does not offer the same quality of service as direct telephone connections, it is expected to expand rapidly in the near future, and it does have many advantages over traditional telephone calls.

Collaboration: Simultaneous collaborative design work at multiple sites; remote control of robotic devices from thousands of miles away via multimedia through satellite or ground-based fiber-optic cable.

Entertainment: Present, market, and sell music, video, etc., online. Users download whatever they want and pay for it.

Video and teleconferencing: Saves travel time and travel cost.

Centralized data access: Keep track of the knowledge bank of your company, and lets everyone use the knowledge gained by others.

The Community

A two-way high-speed connection could be used for interactive applications such as online classrooms, showrooms, or health clinics, in which teacher and student (or customer and salesperson, doctor and patient) can see and hear each other through their computers. An always-on connection could be used to monitor patient health remotely through the Web. Some of the benefits are as follows:

Education: Distance education and training through enhanced videoconferencing with shared collaborative workspace capability.

Scientific applications: Computer reconstruction of human profiles transmitted via broadband networks for simultaneous analysis by police and forensic experts at different locations.

Medical: Patient diagnosis and physician consultations between hospitals in different parts of the country and top medical institutes.

Government services: E-government services such as redress of grievances using multiparty videoconferencing, land record information archives, etc.

Individuals

Broadband access provides users with high-speed Internet access, access to video and music on demand, interactive game playing, and other lifestyle information when they need it. The following are some applications:

Video-on-demand: A study conducted by research firm In-Stat/MDR reports that about one-third of all digital cable TV subscribers in the United States with available VoD service have become regular VOD users. In addition, the firm predicted that worldwide cable-based VoD subscribers would grow from 3 million regular users at the beginning of 2003 to more than 11 million by 2006.

Gaming: Another critical development is the move away from PCs as the predominant device linking to the information highway. Consumer electronic manufacturers are also positioning themselves, mainly through their game machines. All the new devices that are steadily entering the market are broadband enabled, and these can be used for a range of other broadband applications as well.

Personal video recorder (PVR): The PVR allows consumers to record TV programs and replay them when they wish. The service enables consumers to pause, rewind, and instantly replay and play back in slow motion any television broadcast. Viewers can time-shift their favorite television shows and create a customized television lineup for viewing at anytime.

Broadband Technologies

Broadband communications consists of the technologies and equipment required to deliver packet-based digital voice, video, and data services to end users. Today's broadband solutions are quite complex and require semiconductor manufacturers to integrate a wide variety of innovative technologies to offer low-power, cost-effective system solutions that address the needs of original equipment manufacturers (OEMs), service providers, and end users.

There has been impressive growth of the Internet, leading to tremendous buildup of high-speed intercity communications links that connect population centers and Internet service providers' (ISPs) points of presence (PoPs) around the world. This build out of the backbone infrastructure or core network has occurred primarily via optical transport technology.

Until recently, lack of economically viable and sustainable high-speed and wide-bandwidth technology for the last mile was a bottleneck for the connection of homes and small businesses to this infrastructure. Advancements in technology and the falling price of customer-end electronics have made the dream of broadband a reality for homes and small businesses.

Let us understand various broadband infrastructure, access, and home networking technologies and examine the essential technology building blocks required to deliver end-to-end broadband connectivity from the infrastructure to endpoint devices.

ISDN

The most straightforward extensions of telephone network access involve leaving the twisted-pair copper plant in place and digitizing the transport over them. Using basic-rate ISDN, up to 144 kilobits of aggregate bandwidth can be brought to homes and businesses. ISDN technology can thus support two multiuse (voice, data, or limited-speed video) channels to the home and one or more packet data channels. These would enable access to information resources with text and graphics. Basic-rate ISDN falls short of being suitable for full-motion, large-screen video applications. Local telephone companies are beginning to offer basic-rate ISDN for residential consumers, though price packages and ordering processes are complicated, and user awareness and therefore take rates are limited.

Cable Internet Access

As an alternative to existing copper phone wires, cable companies have been providing broadband access by upgrading their cable plant to carry data and voice services in addition to traditional video services.

Today's cable networks generally deliver data with download speeds roughly between 500 kbps and 2 Mbps and upstream speeds of 128 kbps. This data rate far exceeds that of the prevalent 28.8 and 56 kbps telephone modems and the maximum 128 kbps of ISDN, and is about the data rate available to subscribers of Digital Subscriber Line (DSL) telephone service. The actual bandwidth for Internet service over a cable television line is up to 27 Mbps on the download path to the subscriber with about 2.5 Mbps for interactive responses in the opposite direction. However, because the local provider may not be connected to the Internet on a line faster than a T1 at 1.5 Mbps, a more likely data rate will be close to 1.5 Mbps. In addition to the faster data rate, an advantage of cable over telephone Internet access is that it is a continuous connection.

A cable modem termination system (CMTS) communicates with cable modems located at the customer premises to provide broadband access services. All cable modems can receive from and send signals only to the CMTS, but not to other cable modems on the line. Some services have the upstream signals returned by telephone rather than cable, in which case the cable modem is known as a telco-return cable modem.

The cable modem typically provides an Ethernet interface to a PC or to a small router when multiple PCs are connected. A cable modem can be added to or integrated with a set-top box that provides a television set with channels for Internet access. A cable modem has two connections: one to the cable wall outlet and the other to a PC or to a set-top box for a television set.

Although a cable modem does modulate between analog and digital signals, it is a much more complex device than a telephone modem. It can be an external device or it can be integrated within a computer or set-top box. Typically, the cable modem attaches to a standard 10Base-T Ethernet card in the computer.

Newer-generation cable modem technologies will significantly increase the available bandwidth to further enable interactive applications such as videoconferencing and high-end online video. Internet Protocol (IP) telephony is one of the services that can be delivered over coaxial cable. For the cable operators, IP telephony enables them to offer voice services that to date have been the domain of the telephone companies.

Digital Subscriber Line

DSL delivers high-speed Internet access using existing copper telephone lines already installed in hundreds of millions of homes and

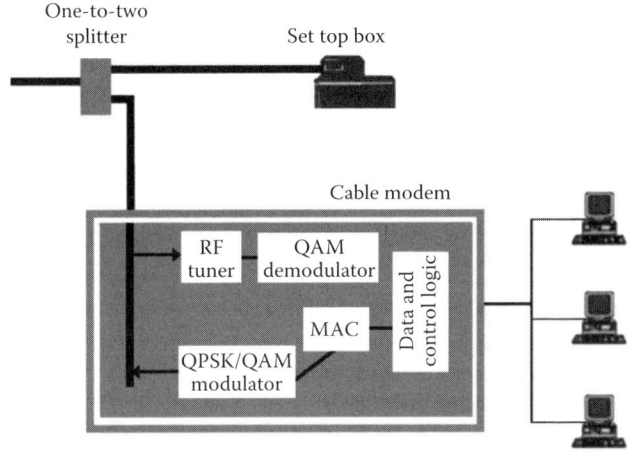

Figure 5.3   Cable Internet access architecture.

businesses worldwide. With DSL, consumers and small businesses get a dedicated, always-on connection to the Internet. DSL provides broadband speeds of up to 8 Mbps, that is, up to 50 times faster than conventional dial-up connections.

The existing copper telephone lines are made up of different bandwidth channels. The lower bandwidth channel carries your voice communication (telephone), which leaves the higher-bandwidth channel available for two-way high-speed data transmission utilizing DSL technology. There is no need for an additional phone line, because DSL uses the higher-bandwidth channel that your telephone does not. Therefore, one can talk on the phone and access the Internet at DSL speed at the same time.

Different variants of DSL exist to address different technology trade-offs that can be made regarding various different network environments and applications. One of the key trade-offs is distance (referred to as reach) from the central office (CO) and another is data rate.

Asymmetrical DSL, or ADSL, is primarily used for residential services. ADSL takes advantage of the fact that there is more cross-talk interference at the CO end of a copper pair than at the subscriber end because of the large bundles of cabling entering the CO. ADSL can provide data rates up to 8 Mbps from the network-to-subscriber direction, and up to 1 Mbps from the subscriber-to-network direction. The asymmetry of ADSL works well for today's home applications, in which the majority of bandwidth is consumed in the network-to-user direction.

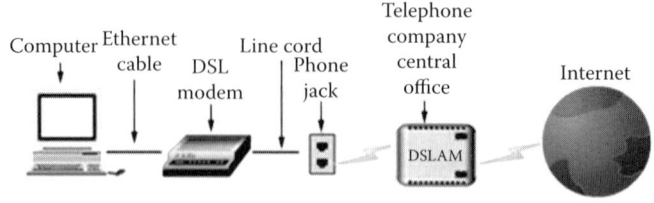

Figure 5.4   Digital subscriber line access architecture.

Symmetrical DSL, or SDSL, is a cost-effective solution for small and medium enterprises, offering a competitive alternative to T1 and E1 lines. The International Telecommunication Union-Telecommunications Standardization Sector (ITU-T) standard G.991.2, also known as G.shdsl, is a replacement standard for proprietary SDSL. G.shdsl offers data rates from 192 kbps to 2.3 Mbps while providing a 30 percent longer reach than SDSL.

Very high bit rate DSL, or VDSL, can support symmetrical or asymmetrical services. Asymmetrical VDSL is capable of providing data rates to the user of up to 52 Mbps, making it suitable for transporting high-speed applications such as real-time video streaming. The trade-off for this high speed is restricted reach. This requires that the customer be located close to the CO or that the infrastructure access gateway reside outside the CO (and closer to the customers) in a remote terminal (RT).

Fiber-to-the-Cabinet and Fiber-to-the-Home

The installation of fiber is being employed for new infrastructure being developed by new service providers or incumbents in areas where copper wires are not currently present. Fiber-optic technology, through local access network architectures such as fiber-to-the-home/building (FTTH/B), fiber-to-the-cabinet (FTTCab), and fiber-to-the-curb (FTTC), offers a mechanism to enable sufficient network bandwidth for the delivery of new services and applications.

A fiber-optic cable is run from the CO to the neighborhood. Passive optical splitters are used to provide point-to-multi-point connectivity. This is referred to as a passive optical network or PON. In the case of FTTCab or FTTC architectures, the signal is converted to provide connectivity to the subscribers via copper pair wires. Because these cabinets are collocated in a neighborhood, the copper pair run is typically less than 3,000 ft, thus enabling high-performance xDSL access to be achieved.

Wireless Access

There are many different wireless technologies that can provide broadband access, and many more are in different stages of development. It is important to choose the technology that best matches the needs of these new markets. Whereas certain technologies have been very successful in urban areas, some are successful in rural areas and others can adapt in multiple environments. These wireless technologies are as follows:

Radio links: Radio communication devices, generally operating in the SW/HF/VHF (1 to 100 MHz) bands, have been very popular in rural regions owing to their ease of use and low-cost, robust technology. Although radio remains the most practical and affordable means of broadcasting and distributing information, the use of radio for two-way communication of digital data has been very limited owing to relatively low bandwidth and a lack of standard hardware. In addition, the design of small, efficient, long-distance antennas at these frequencies is not currently feasible.

Broadband fixed wireless: Provides always-on high-speed access. It is a wireless technology and is ideal for areas that do not have cable or DSL access. It is similar to broadcast television: an antenna mounted on a fixed location signals to the digital receiver on a roof.

Satellite/VSAT: The use of satellite-based services has traditionally been the only alternative in regions where no ICT infrastructure exists. Over the past 20 years, the increasing use of higher satellite frequencies has enabled smaller parabolic antennas and more compact hardware. Although satellite-based connectivity is now becoming more affordable, the hardware cost and service fees are still considerable. To justify the cost of satellite connectivity, a satellite ground station must generally be combined with other wireless network technologies to distribute the available bandwidth and services to a large user base. Other wireless technologies capable of providing broadband services are cellular (GSM/CDMA), Wi-Fi (WAN/LAN), WiMAX, and wireless in local loop. We will discuss all these broadband wireless access (BWA) technologies in detail in the following sections.

Broadband Drivers and Pitfalls

Access to the Internet has become a big business for cable and telecom operators worldwide. But key questions still need to be answered, the answers to which are vital to the future of broadband. What trends will drive broadband in the future? What trends will create barriers to broadband's success in the future?

Catalysts

The higher connection speeds enable multimedia applications such as real-time Internet audio streaming, posting and displaying digital photographs for friends and family, and viewing video clips of news events and movie trailers. Because broadband access is always on, unlike dial-up access, there is no wait to connect to the Internet. Thus, people with broadband access tend to leave their PCs turned on and use the Internet for mundane tasks such as checking television listings and looking up phone numbers, tasks that were not worth the bother when a slow dial-up connection first had to be established.

Another important aspect of broadband access is that it allows people to telecommute effectively by providing a similar environment as when they are physically present in their office: simultaneous telephone and computer access, high-speed Internet and intranet access for e-mail, file sharing, and access to corporate servers.

Broadband is an enabling technology. Its benefits are realized through the delivery of advanced applications and services that are expected to bring about productivity gains both for businesses and public administrations. E-commerce and E-business, for example, become more convenient. They allow business deals to be concluded fast and reshape the supply chain.

Distance education and learning are stimulated through real-time services, resulting in the upgradation of skills, improved human capital, and lifelong learning. In healthcare, high-speed Internet access allows diagnosis and patient treatment to be carried out independently of geographic location.

In the context of E-government, broadband facilitates the online supply of existing and new public services. It improves the efficiency of public administrations and facilitates contacts between citizens and government.

Finally, teleworking, VoD, VoIP, and videoconferencing have become real and practical options. The benefits of broadband play a crucial role in promoting progress toward an inclusive knowledge-based economy and ensure growth through improved competitiveness.

Once broadband access reaches critical mass in terms of market penetration, there will be a new class of end-user devices that will enable many new Internet-enabled applications. Already, people are

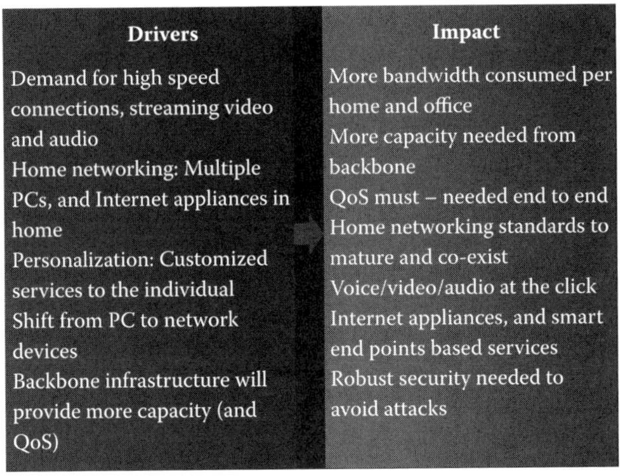

Drivers	Impact
Demand for high speed connections, streaming video and audio	More bandwidth consumed per home and office
Home networking: Multiple PCs, and Internet appliances in home	More capacity needed from backbone
Personalization: Customized services to the individual	QoS must – needed end to end
Shift from PC to network devices	Home networking standards to mature and co-exist
Backbone infrastructure will provide more capacity (and QoS)	Voice/video/audio at the click
	Internet appliances, and smart end points based services
	Robust security needed to avoid attacks

Figure 5.5   Broadband drivers and impact.

able to perform functions remotely via the Internet: monitoring and controlling their homes, viewing their children who are in day-care centers, checking on live traffic conditions, and playing stereo-quality music over Internet radios.

Uncertainties

Although the new developments can be clearly identified, there are still many uncertainties. In any case, we are now in a much better position than before to understand and address these issues.

Not enough is known at this stage about the demand and the acceptability of innovative applications, especially in the home environment. In large parts of the population, there is what market researchers call a *demand gap*. True, the demand for new applications may exist, but only a few have any idea what they are about and what their use may be; therefore, people cannot articulate their requirements.

A multitude of technical developments will enable innovative applications. But, so far the developments that will prevail in the future have not crystallized and, above all, the cost factor is not known.

Broadband has also brought security concerns. For small businesses, cable and DSL service providers typically install simple modems for single-computer always-on Internet access, and they stress the need for security precautions to protect the PC and its resident business-critical data.

The business models — that is, who offers what and who bears the costs — are for the most part still unclear. Also unclear is the framework for political, social, and legal guidelines for the information society that is currently being developed.

Chapter 6

Wired versus Wireless

Wireless is still the IT sector in the telecom landscape.

Wireless voice has been at the top of the growth chart for quite a long time since it displaced the Internet, but history is getting repeated and wireless data connectivity is quickly gaining momentum across the globe. Most home and business users have, to some degree, calculated the advantages of switching data traffic to a wireless connection. Users are becoming more reliant on their mobile phones even for data, thereby using their landlines less and less, and eventually they will see little need to continue both services. Further, the rapid development of broadband wireless access (BWA) technologies has changed the dynamics decisively in favor of wireless.

The boom in wireless networks has meant that in many countries there are now more wireless phone lines than fixed lines. There are a number of reasons for this unexpected boom in wireless networks. Without a doubt, the use of wireless or mobile phones is more convenient and requires less investment than a fixed infrastructure. In addition, a wireless infrastructure has more flexibility than a fixed infrastructure, in which at least the part of the access network closest to the user is dedicated to specific locations, and its profitability depends on the usage. Wireless networks do not suffer from this limitation as they can be shared and reassigned much more easily and can become profitable more rapidly.

Some analysts of the telecommunications industry believe that within a few years, most telephone calls in the residential market will

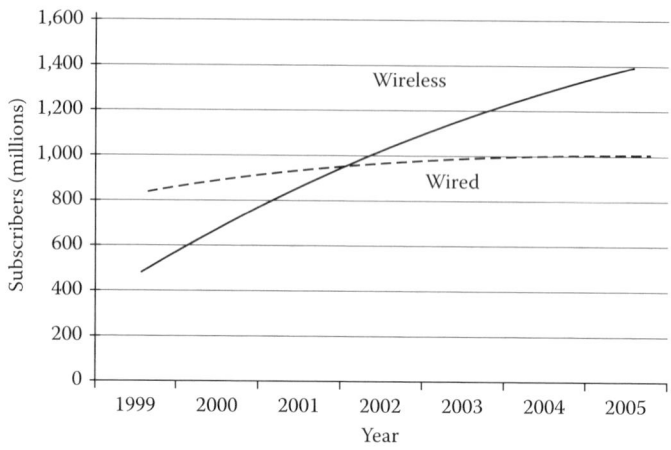

Figure 6.1 Wired versus wireless users.

be placed over wireless networks. These analysts go so far as to suggest that by initiating price wars, the wireless operators could quickly undermine the voice-dependent businesses of most of the landline operators.

The Future of Wireline

Wireless and wireline carriers are aware of this wireline replacement revolution, but are not exactly sure how to respond to it. The wireline carriers are sticking with a "both are best" philosophy, whereas the wireless carriers are encouraging the switch to cellular. This leaves carriers faced with the challenge of offering consumers a technology that bundles all the convenience, ease of use, and functionality of their current landline, broadband, and mobile connections at a price that is less expensive than each separate service.

Although landline operators' profits will take a big hit in the future without any doubt, it would be too audacious and premature to claim that landlines will disappear from the telecommunications landscape.

Any wireless access network eventually gets connected to the wired network, so it is important to understand the role of wireless with regard to the wired parts of a deployment. Certainly, wireless systems have a number of advantages; however, it is important to realize that there is a price to be paid for these inherent advantages.

The primary reason for landline's survival will be its high performance. In the near future, it is unlikely that wireless will deliver the quality and speed of a fiber deployment. Also, recent developments in the area of optics such as optical switching, fiber to home or curb, and extremely high data transmission (tens of thousands of terabits per second of data) bode well for wireline future prospects. Consequently, it is best to see wireless systems as complementing rather than replacing wired ones.

Ideally, fiber would be deployed as deeply into a deployment as is affordable and practical. Whenever new wires need to be deployed to carry data, fiber systems are clearly the right choice. Wireless systems would then be used to extend this connectivity to a larger number of locations and ultimately connect end systems to the network. However, sensible deployments will utilize good copper connections that may already be in place to feed wireless systems, and wireless access systems may well feed locations such as offices that may have an internal wired Ethernet system in place. Thus, most real-world deployments will use both wired and wireless technologies in a cost-effective mix to reach the most number of users.

As a society, we have evolved from radio to television, from black and white photography to color, and from analog to digital networking. The next wave of change is wireline to wireless communications. Let us examine the reasons for this wireless craze.

Why Go Wireless?

The start of the new millennium is witnessing a telecommunications world that is very different from even the recent past. Clearly, there are challenges ahead. But driven by the power of mobility, the world is going wireless, and an ever-increasing number of people everywhere are reaping the rewards of communicating in a world without wires.

The main factor behind this tremendous growth has been the wireless medium's ability to substantially satisfy any two of the three components that comprise the ultimate goal of telecommunication: "any information, anytime, anywhere." A wireless communications system provides anytime, anywhere communications. Some of the inherent characteristics of a wireless communications system that make it attractive for users are as follows:

Mobility: Wireless enables better communication, enhances productivity, and enables better customer service. A wireless communications system allows users to access information and conduct business from anywhere.

Reach: Wireless communications systems mean people are better connected and are reachable wherever they are.

Simplicity: Wireless communications systems are faster and easier to deploy than cabled networks. Installation can take place without hassles, ensuring minimum disruption.

Flexibility: Wireless communications systems provide flexibility as a subscriber can have full control of his or her communication.

Setup cost: The initial costs of implementing a wireless communications system compares favorably to a traditional wireline or cable system. Communications can reach where wiring is infeasible or costly, e.g., rural areas, old buildings, battlefields, vehicles, etc.

Falling services cost: Wireless service pricing is rapidly approaching wireline service pricing.

Global accessibility: Roaming makes the dream of global accessibility a reality as most parts of the globe today are well covered by one wireless services provider or another. Also, roaming service provided by service providers allows flexibility to stay connected anywhere.

Smart: Wireless communications system provides new smart services such as SMS, MMS, etc.

Cultural: A wireless communications system is a personal device, whereas more wireline is more institutional, e.g., associated with an office.

In today's world, wireless communication is no longer just about cell phones; instead, telecommunication seems to be heading toward providing all possible ways to keep information place independent to a lesser (as in the wireless local loop [WLL] case) or greater extent (as exemplified by cellular technologies).

The WLL Revolution

To tackle this issue of getting features mentioned earlier, one significant development that threatens landline operators is the emergence of WLL, a direct substitution of the old copper line. To derive optimum benefit, it is important to define the end goal of wireline replacement, i.e., creating a wireless local loop. The local loop has traditionally referred to the wiring that connects an individual telephone in a residence or business to the central office of the telephone company. To truly refer to a wireless technology as a wireline replacement solution, it should replace all the functionality of a wireline connection — including voice, data, and fax.

Since the advent of the telephone system, copper wire has traditionally provided the link in the local loop between the telephone subscriber and the local exchange. But copper's heyday in the local loop is coming to an end. Economic imperatives and emerging technologies are opening the door for WLL solutions. Sometimes called RITL or FRA, WLL uses wireless technology coupled with line interfaces and other circuitry to complete the last mile between the customer's premises and the exchange equipment; further, it is capable of providing not only toll-quality voice but also video and data transmission.

Best of Both Worlds

WLL is "the hot telecom growth industry of the next decade." The worldwide WLL market has crossed 500 million subscribers at the end of 2005. The vast majority of these lines will be in emerging countries, with a small percentage in developed countries. The two basic market segments for WLL are for basic phone service in emerging economies and for wireless bypass in developed economies.

In developed economies, the relatively low deployment costs, maintenance costs, and learning-curve advantages make WLL a competitive bypass solution and a viable alternative to wireline networks for POTS and data access.

In developing countries, WLL technology has several characteristics that make it attractive to deploy for 20 to 50 percent of a typical telephone network. One important consideration is that a WLL network can be deployed very quickly. This is a key advantage in a market in which multiple service providers are competing for the same user base. In some other cases such as adverse terrain or widely dispersed subscriber areas, WLL would be even more attractive.

Ultimately two issues will determine the growth of WLL: cost and bandwidth. Today's exorbitant access rates, coupled with regulatory changes, have created a competitive environment that gives new operators the incentive to invest in their own WLL networks. As the expense of provisioning service via WLL is not affected by the distance between the subscriber and the central office (CO), WLL is more cost-effective than a wireline operator service provider (OSP). WLL has a much lower incremental investment cost than copper, and it is much cheaper to deploy at lower subscriber densities. The cost of deploying the last mile of connectivity will continue to fall for wireless while

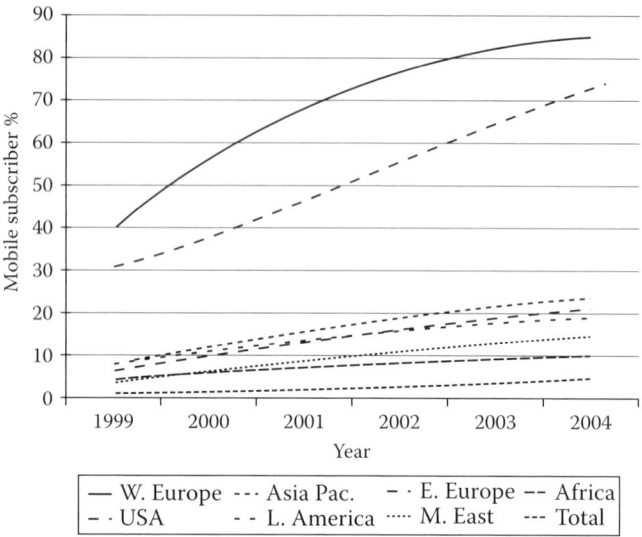

Figure 6.2 Wireless penetration.

remaining constant for copper wire networks. However, WLL deployment costs must be balanced with the potential for lower access fees.

The growing demand for high-bandwidth transmission capable of supporting rich data types places additional requirements on a WLL system. In short, ease, cost, and speed of service implementation and maintenance, along with scalability, reliability, and versatility make WLL the preferred alternative to wireline.

Wireless Data Access

Wireless information delivery has been around for a long time indeed, since the first word spoken by humans. Between 1895 and 1901, Guglielmo Marconi first demonstrated the feasibility and tremendous potential of wireless communications, which was progressing from small homegrown experiments to transmission of the first transatlantic wireless messages.

In the century that followed, our lifestyle and culture was modified dramatically by a series of culture-altering wireless applications, including broadcasting (both audio and video), radar, and mobile telephony.

Yet, we find ourselves once again on the threshold of explosive growth, this time in wireless access for an ever-widening user population. With the development of wireless devices that can log on to the Internet and send and receive data, wireless communication is joining the digital age.

The first wave of wireless data access services is already appearing and will gain momentum because several basic pieces of the solution already exist. Applications and information are in place. Operators have been upgrading the data capabilities of their wireless networks, and gadgets that can access data are available.

Wireless technologies provide mobility, as they allow users to be easily contacted, and also allow them to access information and services regardless of location, time, or the device that they are using. Mobility is truly revolutionizing the way organizations work and is increasingly viewed as a critical factor. With the availability of convergence, a concentration of computer power in a variety of devices such as laptops, dashboard computers, mobile phones, and personal digital assistants has led to an explosion in mobility and, therefore, to higher productivity levels.

Wireless device technology and advanced software, together with the influence of the Internet, are creating a vast array of Web-based services to which people want easy, 24-hr access from wherever they happen to be. Microprocessors are rapidly becoming smaller, faster, more power efficient, and less expensive, which means they will be used more often and in more places to create a host of intelligent devices that will increase access to those services.

Broadband and wireless connectivity are expanding at a rapid rate, providing the final component necessary to create universal connectivity among all of those new devices and the instant availability of the information and services they help deliver.

Connectivity without Strings

One of the next major steps forward in connectivity is to enable users to access the Internet or local area network (LAN) of their institution from a remote location using wireless technology. This type of access is referred to as "wireless and or mobile networks," although there is no commonly accepted term for describing the same. Whereas wireless connectivity has been, until quite recently, limited to relatively slow data rates or to very short access ranges, technology is being introduced that will allow users to connect with data rates that are in excess of home-use broadband connections and with far greater range than is currently available.

In today's mobile business environment, a system that allows users to easily add devices to the network by simply plugging in a wireless

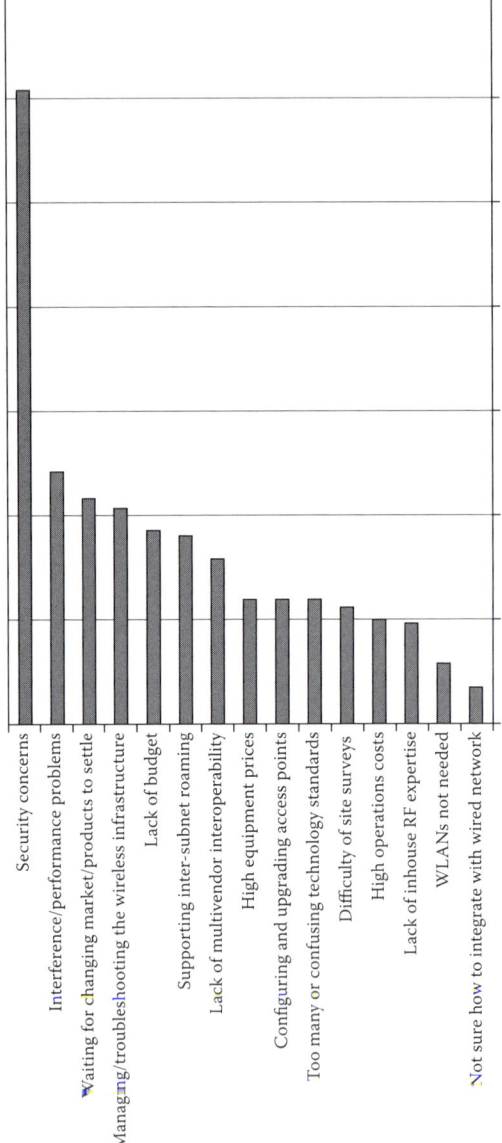

Figure 6.3 Wireless data deployment obstacles.

adapter is extremely appealing. Wireless networking is attractive to businesses because it provides the ultimate in mobility — simple and flexible installation options, low cost of ownership (no cabling costs or maintenance), and excellent scalability.

The potential for increased productivity is enormous as you are connected to your corporate network while in the conference room, at the airport, in the coffee shop, or at home. The affordability and mass appeal of wireless networking is creating further demand for small to mid-sized business network expansion and for broadband as the Internet access medium for networked business.

Wireless Networks

As the name suggests, a wireless network is a network without wires. The tetherless nature of connectivity provides its users almost unrestricted mobility and the means to access the network from anywhere. Whereas in a wired network an address represents a physical location, in a wireless network the addressable unit is a station, which is the destination of a message and is not (necessarily) at a fixed location. Wireless networks operate on a medium that is unprotected from external interference (and thus is significantly less reliable than that for wired networks). Additionally, most of the portable devices are battery operated, and power management is a crucial consideration for wireless devices.

Although wireless networks have been around for some time now, they are fast gaining popularity because of ongoing standardization and the falling costs of hardware components. During the last few years, small to mid-sized businesses have been building and expanding their networks at record rates. Wireless networking, once a technology used mostly by the traveling executive to connect his or her notebook computer to the corporate network when in the office, is playing a larger role today, and Internet access is one of the driving factors.

Internet connectivity is critical to the success of small to mid-sized businesses today. It provides a means of staying in touch with customers and vendors and accessing important decision-making information and research, and it even provides a window through which to keep an eye on the competition. In a typical wireless network infrastructure configuration, there are two basic components:

Access points — An access point or base station connects to a network by means of Ethernet cable. Usually installed in the ceiling, access points receive, buffer, and transmit data between the WLAN and the wired network infrastructure. A campus or building may require several access points to provide complete coverage and allow users to roam seamlessly between access points.

Wireless client adapter — A wireless adapter connects users via an access point to the rest of the network. A wireless adapter can be a PC card in a laptop, an ISA or PCI adapter in a desktop computer, or it can be fully integrated within a handheld device.

The next major steps forward in Internet access are wireless and mobile Internet. In the next section, we will explore and identify the technologies used for wireless Internet connectivity, along with any associated connectivity issues.

Wireless Internet: Boom Time

What a fitting way for the world to usher in the new century, with a breakthrough in communications that provides access to information in all its forms. The convergence of two of the fastest-growing communications technologies ever developed — mobile phones and the Internet — is a potential technological event of the century or a big bang moment for the business world.

Wireless communications and the Internet are already booming, and by the end of 2005 each of these technologies is expected to reach 1.5 billion users; further, there will be more mobile devices connected to the Internet than fixed devices and terminals. The merging of these two complementary paths into a wireless Internet represents the greatest inflection point in the history of communications. For users everywhere, the wireless Internet will mean access to information without restrictions of time or location. For network operators and service providers, it offers a huge array of attractive new business opportunities with solid, continuing revenue streams.

Within the last two years, the mobile Internet market has grown from 200 million to cover more than half of the 1.3 billion mobile phones used worldwide. The 3G system, among other developments, has helped to build the foundations of a cultural revolution whose impact has been compared to the switch from the electric telegraph to the telephone in the last quarter of the nineteenth century.

The number of Internet-connected mobile phones will soon exceed the number of Internet-connected PCs by a wide margin. Worldwide, the number of active PCs is considered between half a billion and three-quarter billion, well shy of that one-and-a-half billion cell phone figure — and the gap is going to grow, particularly in places such as China, where generations may skip the PC altogether and move directly to smaller mobile units of one kind or another.

The PC industry responded early on, making the Internet experience mobile through products such as Centrino-powered laptops, Wi-Fi-enabled pocket PCs, and now WiMAX, which is termed a *disruptive technology*. Although starting slowly, the cellular industry is now well on the way to making its voice-based platform interactive through 3G technology. Both markets are likely to have a bright future in a sector from which different people want different things.

What Is Coming?

Imagine conducting bank transactions, answering e-mail, browsing the Web, and participating in videoconferencing from a single device. When you integrate high-speed data transport, Internet access, and multimedia into one integrated, end-to-end solution, anything is possible.

Although some wireless Internet platforms do indeed provide the user with the same look and feel they are used to on a PC, other devices such as mobile phones, handheld devices, and microlaptops concentrate on sending and receiving timely, relevant snippets of information sourced from the Internet and combined with clever use of subscription information on the tastes and trends of the consumer, both personal and general.

Network providers continuously follow usage trends to determine which wireless Internet services are popular and which are not. They try to develop applications essential to providing customers with reliable and fast service that will work effectively in any format to avoid the kinds of incompatibilities suffered in the early days of the stand-off between Apple and its Mac and Microsoft-Windows-using PC fans.

Benefits of Going Wireless

Businesses today need to be increasingly agile to adapt to changing markets and the changing needs of their customers. With an increasing demand for access to information anytime and anywhere, organizations need to enhance their business processes through effective management of mobile devices and the applications that run on them.

The explosive growth of both the Internet and wireless technology is revolutionizing the way the business is conducted. This trend has created new competitive threats as well as new customer opportunities. Businesses are finding novel ways to add value to their products and services, gain competitive advantage, and increase customer loyalty while also attracting new, high-value clients. There are enormous possibilities that wireless technology can offer.

The challenge, then, is how to turn these possibilities into a reality. The key to success in today's environment is information, whose flow is often impeded only by the need for secured reliable and usable access. It is becoming critical for organizations to enable anywhere, anytime access to information and applications. The main advantages of wireless Internet are described in the following subsections.

Enhanced Customer Satisfaction

By going wireless, organizations have greater flexibility to please customers and support the fundamentals of account management. No longer do customers have to wait until they get to a PC or kiosk, or wait on hold in a telephone queue for information. This convenience not only promises to attract new customers, but also protects existing relationships against aggressive competitors.

Higher Profile in Target Market

Using mobile phones and other wireless devices such as PDAs, customers are just a click away from your Welcome panel, via a branded button that sits on their personal handset screens. Trials show that mobile banking customers tend to interact with their financial institutions more frequently. Every transaction becomes an opportunity to reinforce brand image.

Gain Customer Advantage

Organizations can deliver new services to new customers virtually whenever and wherever they want. With mobile device users growing at an exponential pace, an institution can stand out from the competition by being among the first few to offer the service.

Enhanced Productivity

Organizations with more connected employees have the largest need for wireless data and are adopting enterprise wireless deployments at an increasing rate. The reason is that employees and customers are becoming more and more mobile over time. Regarding this point, the Meta Group states that 80 percent of knowledge workers are mobile for more than 30 percent of their time. Applications that are focused on delivering essential enterprise applications and information such as sales force automation solve a critical problem by maintaining and improving efficiency rates within an enterprise.

Increase Revenue Streams

Services companies need to provide up-to-the-minute information to drive additional revenue-generating activities such as value-added services.

Flavors of Wireless Internet

Wireless communications have become very pervasive. The number of mobile phones and wireless Internet users has increased significantly in recent years. Traditionally, first-generation wireless networks were targeted primarily at voice and data communications occurring at low data rates. The second- and third-generation wireless systems incorporate the features provided for higher data rates. Wireless networks include local, metropolitan, wide, and global areas. These wireless networks have three flavors based on satellite technology, cellular technology, and wireless LAN technology.

Satellite Technology

Satellite technology plays an important role in communication globally. Satellites work by receiving and transmitting radio signals from one earth station to another. Satellite systems have the advantages of transmission from point to multi-point systems, which means transmissions can be beamed to areas that are geographically dispersed. Satellite technology has the potential to beam signals across different countries, which has improved international telephony enormously. It has also improved television signal transmission as programs are transmitted to television operators from one country to another through satellite technology. Over the years, satellite transmission for telephony has been considered inappropriate. This is due to the fact that the time taken to beam the signal to space and back to Earth creates a short delay in the exchange of conversations; this also leads to an echo in telephonic conversations. The fixed cellular system, using satellite

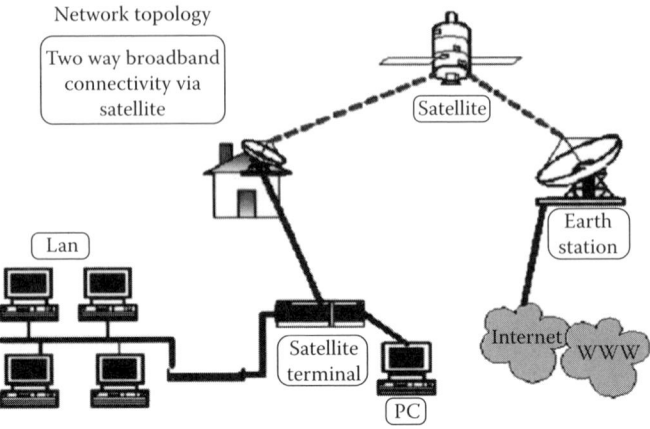

Figure 6.4   Two-way satellite.

technology such as VSAT, has improved access to telecommunication tremendously in Africa.

Cellular Technology

The remarkable growth of cellular mobile telephony, as well as the need for wireless data services, promises an impressive potential for a market that combines General Packet Radio Services (GPRS) and Global System for Mobile Communications (GSM).

Currently, data rates are too slow, and the connection setup takes too long and is rather complicated. Moreover, the service is too expensive for most users. Since the advent of generation-2.5 services such as GPRS, things are far better than they were some time back and look more promising for the future. The introduction of GPRS-to-GSM wireless networks enables the cost-effective and efficient use of GSM networks. Universal Mobile Telecoms Service (UMTS) generation 3 will offer users an alternative in high-speed access, allowing connectivity to the world, from any location on the planet.

First-Generation Mobile Systems

The first generation of analog cellular systems included the Advanced Mobile Telephone System (AMPS) in the United States and Total Access Communications System (TACS) in Europe. AMPS offered 832 channels with a data rate of 10 kbps, whereas TACS was introduced with 1000 channels and a data rate of 8 kbps. Both AMPS and TACS used the frequency modulation (FM) technique for radio transmission. Traffic was multiplexed onto an frequency division multiple access (FDMA) system.

Second-Generation Mobile Systems

Compared to first-generation systems, second-generation (2G) systems use digital multiple-access technology, such as Time Division Multiple Access (TDMA) and Code Division Multiple Access (CDMA). GSM uses TDMA technology to support multiple users.

Examples of second-generation systems are GSM, Cordless Telephone (CT2), Personal Access Communications Systems (PACS), and Digital European Cordless Telephone (DECT). A new design was introduced into the mobile switching center of 2G systems. In particular, the use of base station controllers (BSCs) lightens the load on the mobile switching center (MSC) found in first-generation systems. This design allows the interface between the MSC and BSC to be standardized. Hence, considerable attention was devoted to interoperability and standardization in 2G systems so that carrier could employ different manufacturers for the MSC and BSCs.

In addition to enhancements in MSC design, the mobile-assisted handoff mechanism was introduced. By sensing signals received from adjacent base stations, a mobile unit can trigger a handoff by performing explicit signaling with the network. Second-generation protocols used digital encoding and include GSM, D-AMPS (TDMA), and CDMA (IS-95). The protocols behind 2G networks support voice and some limited data communications, such as fax and short messaging service (SMS), and most 2G protocols offer different levels of encryption and security. Whereas first-generation systems primarily support voice traffic, second-generation systems support voice, paging, data, and fax services.

2.5G Mobile Systems

The move into the 2.5G world began with idea of providing decent data connectivity without substantially changing existing 2G infrastructure. Some of the cellular technologies capable of achieving this goal are considered in the following subsections.

High-Speed Circuit-Switched Data (HSCSD)

HSCSD is designed to allow GSM networks transfer data at rates of up to four times the original network data rates.

General Packet Radio Services

GPRS is a radio technology for GSM networks that adds packet-switching protocols, shorter setup time for ISP connections, and the possibility to charge by the amount of data sent rather than connection time. It is designed to give increased data rates. Also, the charge is based on the amount of data transferred rather than the time spent transferring the data.

The next generation of data heading toward third-generation and personal multimedia environments was built on GPRS and is known as Enhanced Data rate for GSM Evolution (EDGE).

Enhanced Data GSM Environment

EDGE allows GSM operators to use existing GSM radio bands to offer wireless multimedia IP-based services and applications at theoretical maximum speeds of 384 kbps (up to a theoretical maximum of 554 Kbps) with a bit rate of 48 kbps per time slot and up to 69.2 kbps per time slot in good radio conditions. EDGE also let operators function without a 3G license and compete with 3G networks offering similar data services and, in some cases, challenge 3G data rates.

Implementing EDGE is relatively painless and requires comparatively small changes to network hardware and software because it uses the same TDMA frame structure, logic channel, and 200 kHz carrier bandwidth as GSM networks. Designed to coexist with GSM networks and with 3G WCDMA, data rates of up to ATM-like speeds of 2 Mbps could become available (Figure 6.5).

Third-Generation Mobile Systems

Third-generation mobile systems are faced with several challenging technical issues, such as the provision of seamless services across both wired and wireless networks. In Europe there are two evolving networks under investigation: UMTS and Mobile Broadband Service (MBS).

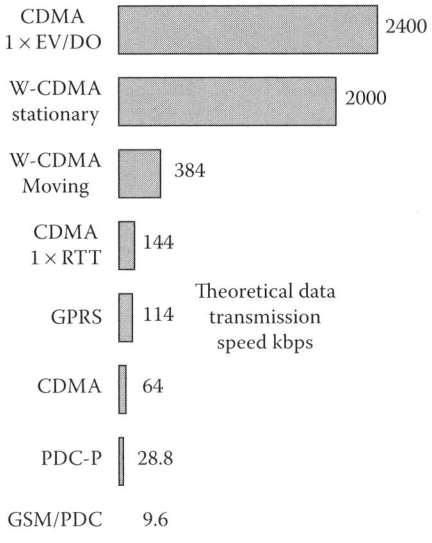

Figure 6.5 Comparative network speeds.

CDMA2000

Growing out of the standard IS-95, CDMA2000 has already undergone considerable development, particularly in the area of multi-channel working. Operators of narrowband CDMA One (IS-95A/B) can deploy services designated 3G in existing as well as new spectrum bands.

Wideband Code Division Multiple Access (WCDMA)

Many see WCDMA technology as the preferred platform for 3G cellular systems as it offers seamless migration for GSM networks which may or may not have already progressed to GPRS/EDGE technology and can provide a migration path for narrowband CDMA networks. Thus, WCDMA will be able to cover much of the world with its comprehensive backward compatibility to such networks.

Cellular telephony evolution from 1G to 3G is depicted in Figure 6.6.

Wireless Network Technology

The huge explosion of wireless technology over the last decade has captured the imagination of technologists around the world. The need

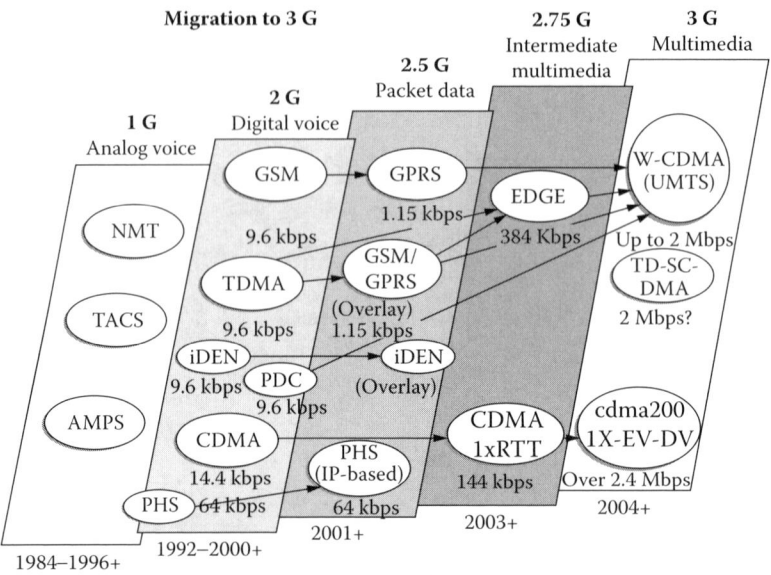

Figure 6.6   Cellular telephony evolutions from 1G to 3G.

for mobility in an ever-changing environment is of paramount importance. From mobiles to laptops and PDAs, the list of wireless technological devices is endless.

There are a number of factors that have made this technology a part and parcel of everyday life. The ability of wireless LANs to provide high data rates at such a low cost, the creation of cheaper components such as chips, and the sheer need for the consumer to have always-on capability are some of them.

It is a human being's fundamental need to communicate, and the telecommunications industry has tapped into an explosive technology that can grow exponentially once the creativity and innovativeness can be sustained. This technology, though in its early stage, has the ability to mature into a very capable, integrated technology. The technologies in this category are as follows.

Fixed Wireless Access (FWA)

As the name suggests, FWA provides fixed wireless broadband access. There are basically two types of FWA systems having different network topologies: point-to-point (P-P) and point-to-multi-point (P-MP). Again, as the name suggests, in P-P a carrier's base station can provide wireless broadband access to a single subscriber, whereas in the case of P-MP, the base station can provide wireless broadband access to multiple subscribers.

Point-to-Point

Current FWA and local services implemented with duplex P-P microwave and millimeter wave systems include fixed versions of mobile cellular systems, competitive local exchange carrier (CLEC) providing broadband wireless access up to 155 Mbps capacity from business premises to long-distance carriers and satellite terminals, private industrial and local government networks, and very small aperture terminal (VSAT) links. Future millimeter wave P-P systems will include a greater variety of narrowband and broadband links from fiber network terminations to CLEC subscribers and private networks.

Point-to-Multi-Point

Current FWA systems such as LMDS or MMDS include simplex broadcast entertainment distribution, simplex educational television, and simplex wireless for television distribution.

Local Multi-Point Distribution Service

LMDS is a high-bandwidth wireless networking service in the 28 to 31 GHz range of the frequency spectrum and has sufficient bandwidth to broadcast all the channels of direct broadcast satellite TV, all of the local over-the-air channels, and high-speed full-duplex data service. The average distance between LMDS transmitters is approximately 1 mi.

Multi-Channel Multi-Point Distribution Service

MMDS operates at lower frequencies, in the 2 GHz licensed frequency bands. MMDS has wider coverage than LMDS (up to 35 mi) but lower throughput rates. Future P-MP fixed services will include the duplex MMDS and 28/31 GHz.

Wireless Local Area Network (WLAN)

This is designed to enable users to access the Internet in localized hot spots via a WLAN access card and a PDA or laptop. Although data

Table 6.1   U.S. and European Wireless Networking Standards

United States	Network	Europe
IEEE 802.20 (proposed)	WAN	3GPP, EDGE
IEEE 802.16 (WiMAX)	MAN	ETSI HIPERMAN, HIPERACCESS
IEEE 802.11 (Wi-Fi)	LAN	ETSI HIPERLAN
IEEE 802.15 (Bluetooth)	PAN	ETSI HIPERPAN

speeds are relatively fast compared to mobile telecommunications technology data rates, their range is short.

Wireless Wide Area Network (WWAN)

This is designed to enable users to access the Internet via a WWAN access card and a PDA or laptop. Although data speeds are very high compared to mobile telecommunications technology data rates, their range is also on the higher side (Table 6.2).

Wireless Personal Area Network (WPAN)

This is designed to enable users to access the Internet via a WPAN access card and a PDA or laptop. Although data speeds are very high compared to mobile telecommunications technology data rates, their range is very short (Table 6.3).

Wireless Region Area Network (WRAN)

This is designed to enable users to access the Internet and multimedia streaming services via a WRAN. Whereas data speeds are very high compared with mobile telecommunications technology data rates as well as other wireless network technology, their range is also quite substantial.

A specific charter of the WRAN working group is to develop standards for cognitive radio-based air interfaces for use by license-exempt devices on a noninterfering basis in spectrum that is allocated to the TV broadcast service.

WRAN, which is presently in its infant stage, is the most recent addition to a growing list of wireless access network acronyms defined by coverage area.

Table 6.2 Different WAN Technologies

Technology	Data Rate	Pros	Cons	Status
GPRS	171.2 kbps	Packet data for the GSM world	Data rates may disappoint	Will be the most successful technology through 2005
HSCSD	115 kbps	Dedicated channels	Low deployment, expensive	Will not become mainstream
EDGE classic	384 kbps	Higher data rates for both packet and circuit	Expensive, little terminal support	Will not be able to compete with WCDMA
EDGE compact	250 kbps	Higher data rates for both packet and circuit TDMA networks	AT&T (main proponent) has changed direction	Unlikely to be successful
CDMA/IS-95B	115 kbps	Interim packet tech for CDMA networks, backward compatible with IS-95A	Only adopted in Japan and South Korea	Most carriers will prefer to deploy CDMA2000 1×MC
CDMA2000 1×MC	307 kbps	High data rates, smooth migration path	Limited global footprint	Good technology but will not survive
PDC-P	9.6 kbps	Used by NTT DoCoMo	Japan only, low data rate	Currently the most successful wireless packet technology in the world

Table 6.2   Different WAN Technologies (continued)

Technology	Data Rate	Pros	Cons	Status
W-CDMA	2 Mbps	Massive industry support	High license fee	De facto global standard
CDMA2000 3×MC	2 Mbps	Backward compatible with 1×MC and IS-95A	Support has cooled down	Good technology but unlikely to succeed
CDMA 1 EVDV	2.4 Mbps	Smooth migration path	Limited global footprint	Will not become mainstream
CDMA 1 XTREME	5.2 Mbps	Very high data rates	Proprietary — Motorola, Nokia	No indication of intent from carriers

Table 6.3   Different PAN Technologies

Technology	Data Rate	Pros	Cons	Status
Bluetooth	723.2 kbps	Low cost	Interference, security	Replace cables
Infrared	115 kbps	Very low cost	LOS	Replaced by Bluetooth
802.15.1	723.2 kbps	Low cost	Interference, security	Formalized Bluetooth
802.15.3	>20 Mbps	High data rates	Expensive, not backward compatible	Unproven business case
Ultrawideband (UWB)	>20 Mbps	High data rates, no dedicated frequency	Not approved, expensive	Underhyped, potentially disruptive, launch by 2006

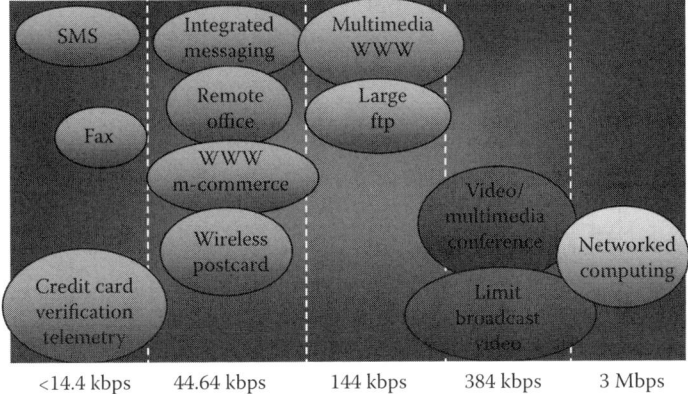

Figure 6.7   Wireless data speed and applications.

Wireless Networks and Mobile Convergence

This refers to access technology that allows users and devices to swap between telecommunications technologies (e.g., GPRS, 3G) and public access WLAN to gain the highest available data rate, depending on their geographic location. It is now being discussed, and it is likely that devices that allow seamless roaming between technologies will become available.

Drivers for Wireless Networks

It is by improving business processes that wireless access will find its place in many enterprises. Several internal and external factors are converging to create a sense of urgency in businesses to find these process efficiencies:

   Increased customer expectations: With the progress of the Internet era, customers expect instant service and problem resolution.
   Need for effective time utilization: Employees commute and travel more extensively, cooperate across time zones, and have to cope with increasing workloads.
   Need for employee empowerment: Employees need to make informed decisions and act on them anytime, anywhere in the face of more time spent away from the office.

Table 6.4 Different Wireless Access Technologies

Technology	Standard	Usage	Throughput	Range	Frequency
UWB	802.15.3a	WPAN	110–480 Mbps	Up to 30 ft	7.5 Ghz
Wi-Fi	802.11a	WLAN	Up to 54 Mbps	Up to 300 ft	5 Ghz
Wi-Fi	802.11b	WLAN	Up to 11 Mbps	Up to 300 ft	2.4 Ghz
Wi-Fi	802.11g	WLAN	Up to 54 Mbps	Up to 300 ft	2.4 Ghz
WiMAX	802.16d	WMAN	Up to 75 Mbps (20 Mhz bandwidth)	Typical 4–6 mi	Sub-11-Ghz
WiMAX	802.16e	Mobile WMAN	Up to 30 Mbps (10 Mhz bandwidth)	Typical 1–3 mi	2–6 Ghz
WCDMA/UMTS	3G	WWAN	Up to 2 Mbps (Up to 10 Mbps with HSDPA technology)	Typical 1–5 mi	1800, 1900, 2100 Mhz
CDMA2000 1 × EV-DO	3G	WWAN	Up to 2.4 Mbps (typical 300–600 kbps)	Typical 1–5 mi	400, 800, 900, 1700, 1800, 1900, 2100 Mhz
Edge	2.5G	WWAN	Up to 348 kbps	Typical 1–5 mi	1900 Mhz

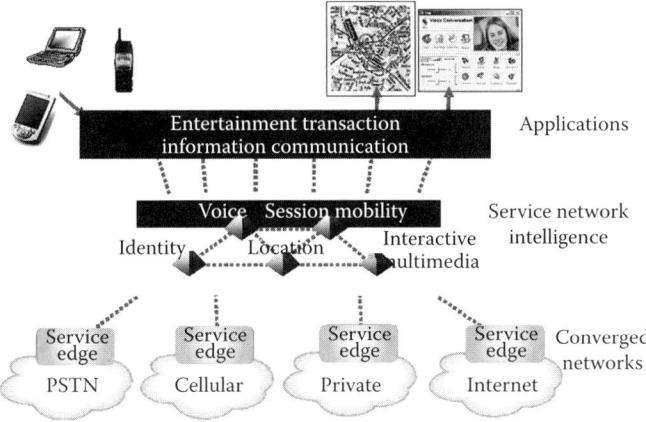

Figure 6.8 Providing value to the end user.

Cost reduction and cost avoidance: Enterprises aim to shorten business process cycles through reduced manual workflow and data reentry and errors to keep a minimal cost base.

Advancing enterprise connectivity: Business requirements to connect processes across the entire value chain are growing as enterprises interact electronically with both internal and external units (suppliers, buyers, and customers).

Technological progress: VPNs, wireless broadband access, and higher wireless security increasingly enable mobile connectivity and help enterprises stay on the competitive edge.

Legislation and government requirements: Insurance, education, social services, and law enforcement agencies are all subject to new and changing government mandates to document activities, improve public service, and share information across geographic boundaries and departments. Mobility provides tangible benefits in terms of cost savings and new revenue as well as intangible benefits such as better customer service and higher job satisfaction for employees.

Issues for Wireless Networks

As with any relatively new technology, there are many issues that affect implementation and utilization of wireless networks. Many challenges are yet to be overcome. There are both common and specific issues depending on the type of wireless network. Some of the common factors include electromagnetic interference and physical obstacles that limit coverage of wireless networks, whereas others are more specific, such as standards, data security, throughput, ease of use, etc.

Standards

A major obstacle to the deployment of wireless networks is the existence of multiple standards. Whereas GSM is the only widely supported standard in Europe and Asia, multiple standards are in use in the United States. As a result, the United States has lagged in wireless networks deployment. Just recently, organizations have been formed to ensure network and device interoperability. For example, the adoption of 802.11 and 802.16 standards have made wireless data networks one of the hottest newcomers in the current wireless market.

Coverage

Another issue is coverage. Coverage mainly depends on the output power of the transmitter (which is generally regulated), its location, and the frequency used to transmit data. For example, lower frequencies are more forgiving when it comes to physical obstacles (walls, stairways, etc.), whereas high frequencies require clear line of sight. For each particular application, throughput decreases as the distance from the transmitter or access point increases.

Security

Data security is a major issue for wireless because of the nature of the transmission mechanism (electromagnetic signals passing through air). It is commonly believed that voice applications are less secure than data applications. This is due to the limited capabilities of existing technologies to protect information that is being transmitted. For example, in metropolitan areas, users are at risk of simple scanning devices hijacking cell phone numbers for malicious use. In WLANs, authentication and encryption provide data security.

Interoperability

Interoperability of wireless networks is a key design objective to drive solution costs down. Otherwise, wireless solutions will only create islands of costly proprietary networks.

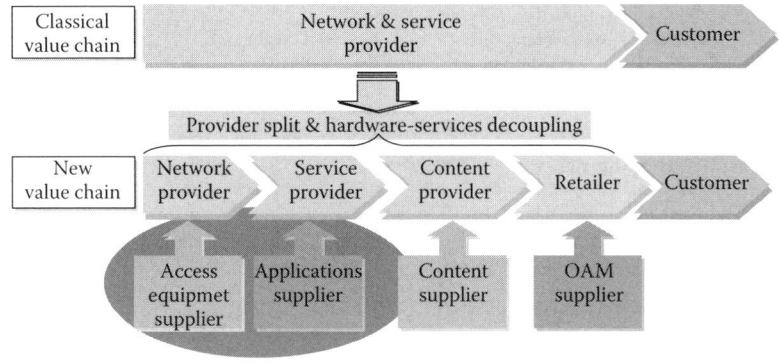

Figure 6.9 Changing business dynamics.

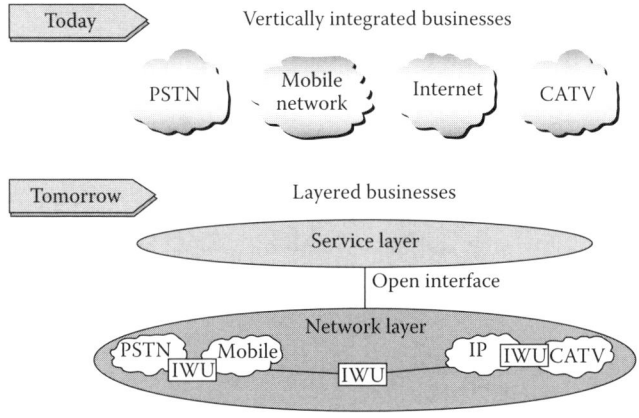

Figure 6.10 Changing business model.

Models

There has been a paradigm shift in business models, following the emergence of wireless systems. The value chain has altered dramatically. There has been significant unbundling of services. No more are access service providers seen as application providers, and so on and so forth (Figure 6.10).

Similar changes also will be seen in future, with players consolidating in their specific areas of strength. New players such as virtual network operators with strong customer-side competencies such as marketing and branding will emerge. They will be served by a real network operator not involved in customer-side issues.

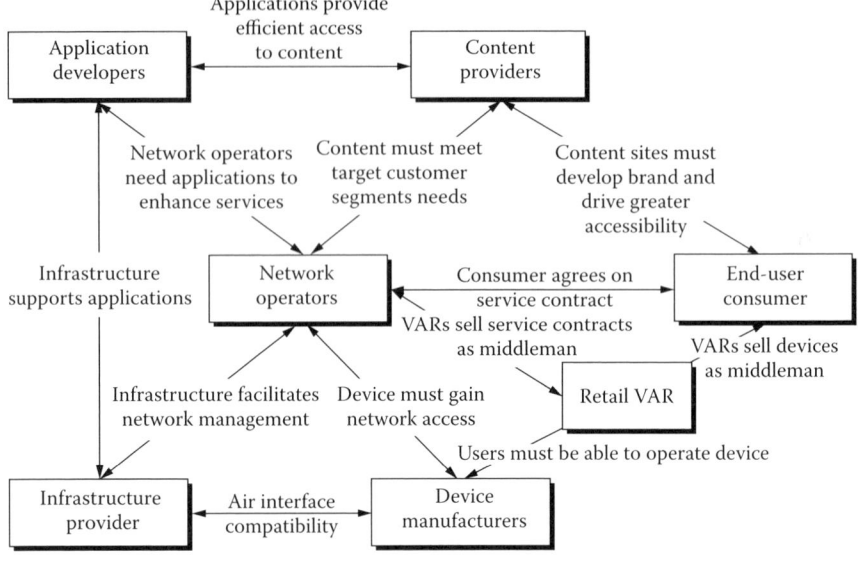

Figure 6.11   Wireless market model.

More actors will set foot in the already fiercely competitive markets, making costs go down while quality of services as well as type of services will increase (Figure 6.11).

Cases and Examples

Here are a few examples of applications the wireless Internet will make possible:

- Enterprise VPNs will extend corporate intranets to remote offices, mobile users, and telecommuters. For example, a salesman traveling in a remote location will be able to check his corporate intranet for product availability without the need for wires.
- Unified messaging via Web-based wireless access will let cellular phone users simultaneously send and receive voicemail, e-mail, and fax messages while talking on the same phone. No longer will users need multiple voice and e-mail boxes for work and home, as it will be possible to consolidate all into one Web-based box.

■ Wireless commerce through the use of smart cards embedded in wireless devices will enable users to make real-time transactions such as buying and selling stocks, purchasing airline tickets, or buying the latest best-selling book — all while simultaneously talking and surfing the Web. News-on-demand will enable publishers and broadcasters to deliver data, audio, and video stories to subscribers on demand, whether it is a cable sports channel or a local broadcast of a football match.

Chapter 7

Broadband Unwired

> It [the rise of wireless data] is going to be disruptive to the people who don't take advantage of it. Entire vertical industries like construction and retail are going to be changed by broadband wireless.

> **Sean Maloney, Intel**

Access to efficient broadband Internet connectivity makes many tasks faster and easier. Wireless local area networks (WLANs) now offer high-speed Internet access at numerous locations in both public and private environments. Wireless broadband access to the Internet has recently witnessed explosive growth. Much of this growth has come from the rise of WLANs. Today, they are being widely used in markets such as education, healthcare, manufacturing, retail, hospitality, government, and transportation. Wireless networking makes access to broadband connections efficient and inexpensive as well as omnipresent.

As many see it, the local loop — the first mile from a user's point of view — has been a bottleneck to networked communications. The relatively limited performance of hundreds of millions of users' access lines worldwide stands in sharp contrast to the high performance of equipment at the ends of those lines.

Except for bandwidth limitations in the local loop, current data networking technology promises to deliver motion pictures and other high-bandwidth material. And the Internet has proved that people love data connectivity. Data networking traffic volume already is comparable to that of the global voice network, with most of its growth coming in the mid-1990s, as the Internet caught on among consumers and businesses when browsers and other software made the Web accessible.

Although the number of Internet users has more than doubled during the past few years, the number of Web pages such users can visit has increased ten times. Optical core networks have supported this growth. Increased bandwidth in the last mile is the primary requirement for achieving the benefits of Internet growth on a wide scale.

The Last Mile Shall Be the First

What is hot in communication today is the high-performance broadband network connectivity for the last mile. The last mile is actually today's best hope for dramatically boosting networking performance.

Digital electronic systems' performance has increased rapidly and steadily, as semiconductors double in capability approximately every 18 months. From personal computers in a home to switches or routers in a service provider's network, every piece of electronics has improved drastically over the time. Furthermore, the capability of information transport equipment is skyrocketing, thanks to optical technology, in which the doubling of capability occurs even more frequently than it does for microchips.

In contrast, the key economic challenge is still how to widen the aforementioned bottlenecks on existing lines on a grand scale (for the world's 750 million access lines) with dramatically higher-bandwidth performance?

So what is the best solution to the bandwidth bottleneck in the last mile? It depends on several considerations: for example, what access plan is already there (if any is there at all), the cost reductions to come, emerging technologies, performance questions that will not be answered until deployments yield real-world data, and the new services that end users demand. One certainty cannot be disputed, though: the outcome of the networking revolution has much to do with discovering and deploying new forms of access.

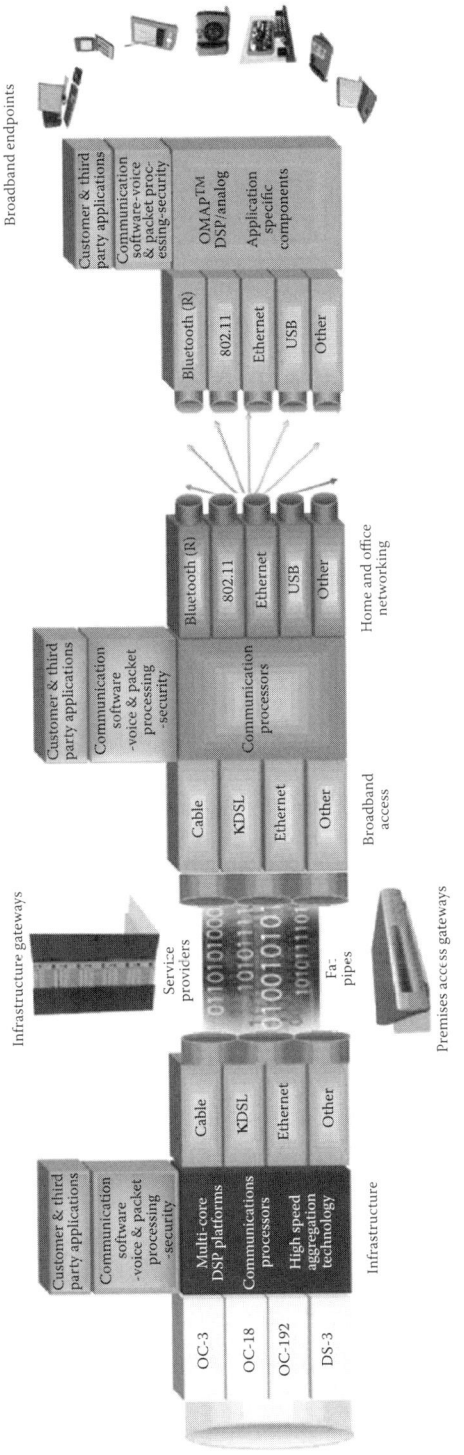

Figure 7.1   Broadband wireless access — architecture.

What Is Access?

Access is what happens in the last mile between a network and an end user or endpoint. Access is fundamental in communications. In its most general sense, *access* means "the ability to make use of" or "the action of going to." Literature sometimes refers to the term *access line*, which typically is defined as "the connection to a local switching system for access to a network."

Access later acquired a technical meaning in the context of subscriber (or digital) loop carriers, meaning "concentrating a set of user connections, prior to presenting them to the switch." Currently, access is generally defined in one of two ways: the first point in a network that an end user's call reaches before it is switched, or getting a service — voice, data, or video — into and out of a network.

Today, access products face the technical challenge of providing "an on-ramp to a network as wide as the lanes of the highway it meets," i.e., in general terms, a broad access path into a network. As networks merge voice with the data infrastructure, converging network models will shape evolving access techniques, but they may also threaten simplicity and economy.

Optical, semiconductor, and wireless technologies can provide the basic resources for widening the bottlenecks. Wireless, fiber, and copper (any and all media) today figure in many technology advancements in this direction. Each medium has its role, each with its own limitations and strengths, to traverse the first (and last) mile between the user and network. Having said that, we also accept that there is a "first among equals."

Communications Revolution: Broadband Wireless Access (BWA)

> BWA is access to broadband communications without any physical connection to a network.

Free space supports the access of mobile cellular and nonmobile wireless customers. Demand for both wireless data services and voice-based services on wireless data is growing even faster. The technology is now available to serve such demand. Further disruptive advancements are just around the corner.

As we march ahead in the new millennium, a fresh breed of wireless technology is taking shape. In hotels, coffee shops, and airports lounges around the world, access nodes for WLAN-based Internet connectivity are coming up. A number of community wireless networks have also come up in Europe, North America, and Australia.

Although this technology offers the flexibility of building ad hoc networks of computing resources in a community, it also opens up additional avenues for building a wireless-based last mile in areas where wireline connectivity is not feasible. These networks also support some bandwidth-sensitive applications such as streaming video, Voice-over-IP (VoIP), and audio- and videoconferencing.

Wireless technology enables operators to deploy service quickly through the installation of inconspicuous radio solutions and to do so incrementally as demand grows. A telecommunications operator can use wireless broadband to develop a customer base to the point where a locale justifies investment in fiber. Wireless broadband services, which are based on a pay-as-you-grow model, are a cost-effective alternative to fiber or copper because deployment and fixed costs are relatively low.

BWA: It's Different

BWA is different from whatever we have seen so far in communication.
Unlike narrowband wireless:

Multi-Mbps, not limited to a few 100 kbps
Always on, bandwidth on demand, not circuit oriented

Unlike broadband wireline:

Fading, interference, multi-path, non-line-of-sight (NLOS) and obstructed-line-of-sight (OLOS) conditions
Limited bandwidth available

Unlike LAN:

Scalable to hundreds of users
Spectral efficiency is the key
Quality-of-service requirements — not just best effort

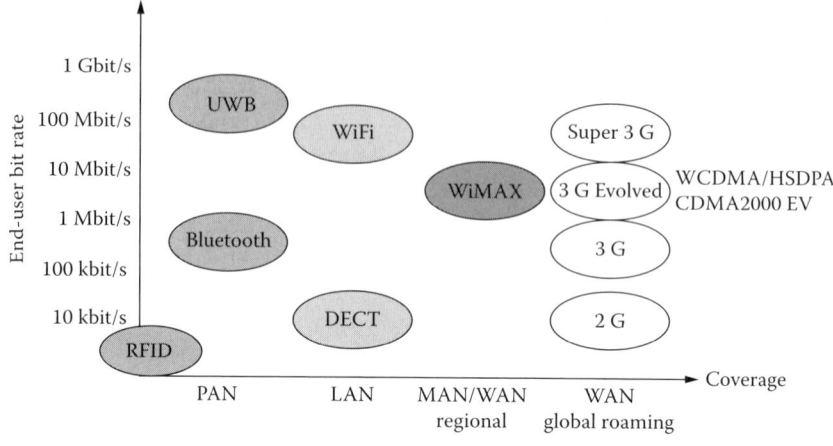

Figure 7.2   Broadband wireless access technologies.

BWA: Why the Hype?

BWA technologies (Figure 7.2) have implicit advantages such as the ability to serve the customer everywhere and mitigate the digital divide, offer right capacity at the right cost, and allow the customer to remain connected while on the move.

BWA is scalable in capacity and coverage areas, an imperative as requested services evolve toward increasingly bandwidth-hungry applications. BWA promises high-speed data, voice, and video services.

BWA offers a last mile connection, as many customers are outside the range of Digital Subscriber Line's (DSL's) or broadband cable's reach, so that these barriers can be lifted and new customers can be captured.

BWA offers faster time to market, and lower total cost of ownership. Further, it is faster to deploy and more flexible; thus, it gives an alternative service to customers who are not satisfied by their wired broadband.

A lot of developments are taking place in this space, with constantly evolving applications, technologies, business models, and regulatory environment. BWA promises an exciting journey to the future.

BWA Technologies

Opportunities for wireless broadband abound in the developed world, but more promising are the countries with relatively underdeveloped network infrastructure, where cost is critical. In many such cases, BWA can facilitate access to the information superhighway quickly and easily in any locale.

Governments around the world are opening up the radio spectrum to make possible new services through wireless broadband technology, which operates in the 10 GHz to 42 GHz range, one order of magnitude higher than mobile wireless frequencies. Developers are confident that atmospheric interference, such as rain, will not limit the potential of this technology to deliver broadband access to users at concentrated endpoints.

Wireless Local Loop

Depending on whom you talk to, the term *wireless local loop* (WLL) can mean any of a number of different things. It most commonly denotes the use of wireless radio signals to provide either voice or both voice and data services to fixed-point subscribers (primarily residential) who are not currently served by landlines.

In many cases, WLL access is delivered via the Code Division Multiple Access (CDMA) protocol (commonly used for cellular telephone service). CDMA is a well-developed protocol, and in regions where CDMA is used for cellular service, access to the network is extremely prevalent.

The basic WLL implementation involves setting up a number of microcells — transmitters that can send and receive data — that connect back to a wired connectivity source such as a phone center or broadband access point. Individual subscribers receive wireless devices to connect to their computer that access the network via the WLL microcells.

Wireless IP Local Loop

Wireless IP local loop (WipLL; Figure 7.3), is a non-standards-based wireless broadband delivery system that supports both voice and high-speed data. The range for WipLL devices varies from 4 to 17 mi with data rates up to 4 Mbps. WipLL technologies is in many ways similar to the forthcoming 802.16 technologies but is strictly a proprietary solution. WipLL also features real-time adaptive modulation (2-, 4-, and 8-level frequency shift keying [FSK]) and automatic repeat request (ARQ), such as 802.16. These features offer high-quality services while maximizing spectrum utilization.

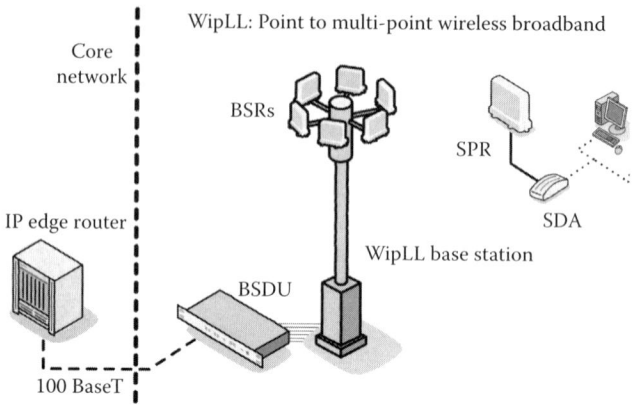

Figure 7.3   Wireless IP local loop (WipLL).

WipLL systems are designed to offer broadband access to a wide range of customers. The heart of each system is the base station (BS), which provides radio access for the subscriber terminals (ST) deployed at the end users' location. The BS connects to the IP cloud using industry-standard 100BaseT Ethernet interfaces. WipLL supports two main types of STs — (1) a split version comprising a compact outdoor unit and an indoor unit and (2) an all-in-one indoor unit.

WipLL is ideally suited to incumbent local exchange carriers (ILECs), competitive local exchange carriers (CLECs), Internet service providers (ISPs) and enterprises wishing to roll out high-speed, high-quality IP-based services to high-end residential, small office home office (SOHO), and business users. A WipLL solution will vary depending on the vendor, but the basic concept is similar for all implementations. WipLL networks are IP based with a proprietary air interface and Ethernet connections for end users (subscribers).

Local Multi-Point Distribution System (LMDS)

LMDS represents an LOS, fixed wireless broadband access technology that operates in several frequency bands in the 28 to 31 GHz range. The bandwidth allocated to LMDS in the United States is either 150 or 1150 MHz, which is by far the largest ever allocated to a wireless transmission method. LMDS originally was viewed as a mechanism to provide a wireless cable TV (CATV) supplement to compete with existing CATV systems. However, its large bandwidth also makes it suitable for high-speed data transmission that can compete with other last mile access technologies. These technologies include cable modems, DSL, and conventional T1 and T3 transmission facilities used for 1.544 Mbps and 45 Mbps access to the Internet, Frame Relay networks, and corporate networks.

LMDS is a wireless system that employs cellular-like design and reuse, except that there is no handoff. It can be argued that LMDS is another variant of the WLL portfolio referenced as proprietary radio systems. LMDS can be a very cost-effective alternative for a CLEC. With LMDS, a CLEC can deploy a wireless system without the heavy capital requirements of laying down cable or copper to reach customers. The cost-effectiveness arises out of the capability to focus the capital infrastructure where the customers are and, at the same time, deploy the system in an extremely short period.

802.11x Wireless

Growth in WLANs can be traced to the creation of 802.11, the IEEE technical standard that enabled high-speed mobile interconnectivity. After sustained efforts by the IEEE 802.11 WLAN Standards Working Group, the IEEE ratified a new rate standard for WLANs, 802.11b, also known as Wi-Fi (Wireless Fidelity).

This standard was certified by the Wireless Ethernet Compatibility Alliance (WECA). The 802.11a standard — approved by the IEEE at the same time as 802.11b — provides for data rates of up to 54 Mbps at 5 GHz frequency. The 802.11g standard, with an even higher data rate, was recently introduced, and operates on the same frequency as 802.11b. Of all these emerging standards, 802.11b has been the most widely deployed, and our subsequent discussion on Wi-Fi will mainly refer to this standard.

802.15x Ultrawideband (UWB) Wireless Networks

UWB may well be the technology that helps realize the dream of the digital consumer and the "connected home." UWB promises to deliver the bandwidth and quality of service that many consumer electronics companies are looking for. A group of UWB companies — the UWB Multi-Band Coalition — is currently spearheading the standardization of the ideal UWB technology for IEEE 802.15.3a. UWB is likely to be first used in consumer applications within the home, with several companies already using the technology to develop applications allowing DVD-quality video content to be streamed.

802.16x (WiMAX) Wireless Networks

The IEEE 802.16 wireless network protocols are the next evolution of the 802.x standards that currently contain the rest of wireless networking technologies standards such as the Wi-Fi and Bluetooth protocols. Products based on the 802.16 protocol will enable transmission of broadband connectivity from a city to outlying villages.

The 802.16x IEEE standards define wireless networking protocols geared toward metropolitan area networks with a range of approximately 31 mi. The standards are in varying stages of development; 802.16a was ratified in January 2003, and 802.16c is on the verge of being ratified. The 802.16a standard holds a lot of promise for rural and developing-world scenarios, with non-LOS connectivity and enough bandwidth for most foreseeable applications.

802.16a-based networks do not have an LOS requirement; however, non-LOS implementations may experience lower bandwidth beyond the range of 4 to 7 mi. It is highly likely that some 802.16x products will have the capability to be chained together to extend the 50 km range; there are technical issues relating to signal overlap and interference that will need to be worked out, but this should be a feasible scenario.

802.20 Mobile Broadband Wireless Access (MBWA)

The IEEE 802.20 Working Group is in the process of developing a new wireless networking standard for MBWA. The 802.20 standard will define the PHY and MAC layers for a high-bandwidth, IP-based, fully mobile wireless network. The group's intention is to fill the gap between existing 802 standards with high data rates and low mobility and existing cellular standards with low data rates and high mobility.

802.22 Wireless Regional Area Networks

The IEEE has a new working group up and running, the IEEE 802.22 with a new acronym, WRAN, which stands for wireless regional area network. WRAN will attempt to bring order to new unlicensed UHF/VHF bands when they are opened as a part of the mandated digital television upgrade. The working group's charter is to develop a standard for a cognitive radio-based PHY and MAC layers of air interface for use by license-exempt devices on a noninterfering basis in a spectrum that is allocated to the TV broadcast service.

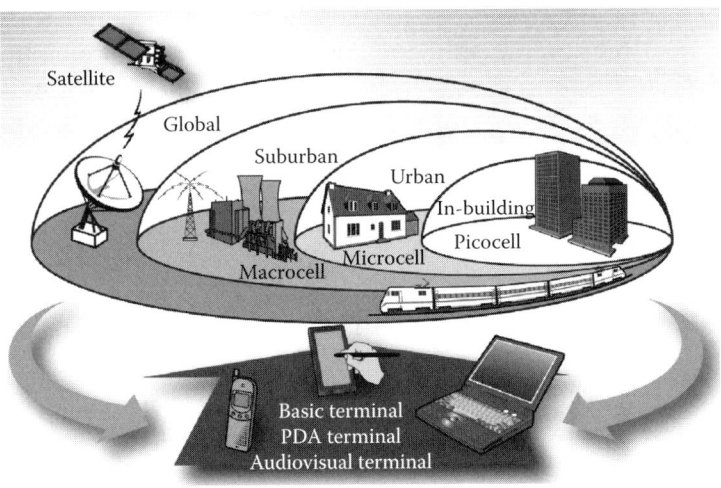

Figure 7.4   Connected anywhere.

Mesh Networks

Mesh networks are wireless data networks composed of two or more autonomous, self-organizing nodes. The nodes are similar to traditional wireless transmitter receivers (akin to a BS or wireless network card) but have additional intelligence built in that enables them to act as minirouters for the network. By adding the capacity for each node to route packets to other nodes in the network, meshes can extend the range of wireless technologies such as 802.11b/g and to provide low-cost coverage of a geographic area using a single broadband connection. (Figure 7.4)

Trends and Directions

Traditional broadband services (DS1, DS3, OC-3, etc.) delivered via wired (copper and fiber) network infrastructures are now being offered via wireless technologies. High demand for such services and deregulation of the communications business is accelerating the development of wireless broadband technologies as CLECs and inter exchange carriers (IECs) move to offer broadband services using these technologies.

Early adopters of broadband wireless have used the technology to replace or replicate mobile communications. Although cell phones have enjoyed ubiquity for a long time, handheld devices in locations ranging from the factory floor to exchange trading floors and sporting events are a more recent phenomenon. The process of adapting both the application and the user in a more like-for-like environment has been relatively painless and is often seen as an enhancement to an organization's functionality, rather than cutting-edge wholesale change.

Until recently, no one was betting on BWA as a disruptive technology, as was the case with mobile or cellular phones. Although a few pioneers implemented a series of wireless technologies with mixed results, the rest adopted a wait-and-watch policy for the adoption of what they frequently referred to as a "better mousetrap" and nothing more.

Things changed drastically in the past two years primarily due to two reasons, which had a cyclical effect. First, because of a fall in the price of equipment (drastic reductions in the cost of wireless microchip), technology became commercially viable at that price point, which led to an increase in the number of units sold. The increase in volumes led to further price reduction, and hence further enhanced volumes. Second, new developments in technology, both in applications and the equipment side, led to an unprecedented interest in wireless.

The wireless networking technologies based on IEEE 802.11 and 802.16 standards, collectively called BWA, have emerged as digital communications standards over the past couple of years, producing a dynamic value chain of suppliers, vendors, and consumers. The technology's affordability, ease of setup, standardization, and favorable regulatory environment in target markets have enabled rapid development, with some clear cost and performance advantages over wireline networking (i.e., Ethernet), for homes and offices. It has reduced total cost of ownership for networking and served as a high-performance medium to distribute available bandwidth.

The wireless broadband market forecast is as follows:

- Ten million BWA customers worldwide by 2008.
- WiMAX product sales will reach $1 billion by 2008.
- The market for long-range wireless products based on 802.16 and the forthcoming 802.20 standard will reach $1.5 billion by 2008 (Figure 7.5).

Successes and Failures

The telecommunications market, because of its inherent nature, has plenty of potential. But examples such as Wireless Access Protocol

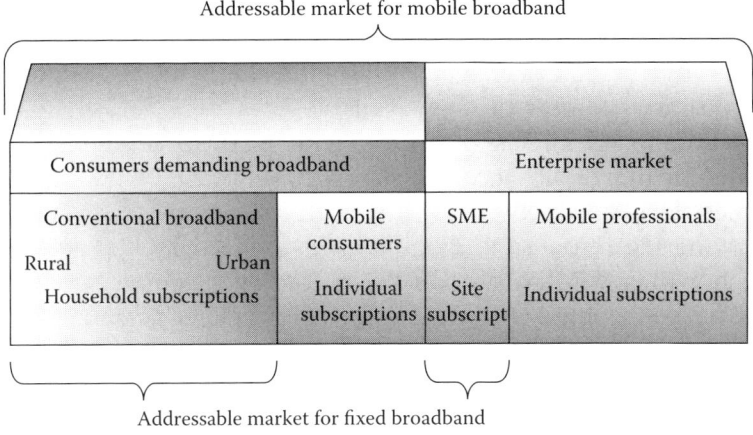

Figure 7.5   **Addressable market for mobile broadband.**

(WAP), UMTS, and mobile data (so far) show that very often the industry does not find returns, because it invests heavily in technology and development that there is no clear customer for. This story seems now to be repeating itself for broadband wireless. Broadband wireless might be a golden nugget, but the opportunity is foggy and investing heavily in developing only the technology will not help dispel that fog.

Today, telecommunications subscribers are not satisfied by simple voice services, they are looking for sophisticated services that can enhance their lifestyle and provide entertainment, as well as communication. That is the good news for wireline and wireless operators, who need new sources of revenue to offset the competitive market dynamic that is driving revenues derived from voice usage to a new low.

To take advantage of today's most promising new source of revenue, service providers must confront two key challenges regarding service delivery:

- First, they must find a cost-effective way to supply adequate bandwidth for an array of multimedia capabilities, whenever subscribers demand these features.
- Second, they must be able to overcome the complexity of the current global infrastructure, in which a broad assortment of technologies and end-user devices coexist.

Amid this diversity, wireline and wireless operators need to transport the same set of services seamlessly across any technology and environment to deliver them and a common end-user experience to any type of device — phone, PC, laptop, or personal digital assistant (PDA). The need of the hour is an Internet access solution that can be accessed anytime, anywhere at the speed of thought, simply by plugging in one of the various end-user devices available.

New broadband wireless technologies are ready to satisfy these needs, supported by universally accepted standards set by the IEEE. The success of wireless local area networking, the 802.11 standards or Wi-Fi, and the initial promising response to long-distance connectivity, the 802.16 standards known as WiMAX, support the preceding statement.

The unprecedented current popularity and meteoric rise of wireless networks is definitely one of the fastest in the technology universe, but it was no overnight success. It has been more than a decade since the IEEE 802.11 effort began. Today, many technology and business practitioners alike are drowning in the alphabet soup of standards, but few who try wireless networks voluntarily give it up. For all its complexity, the appeal of wireless networks is undeniable, and we are now seeing the expansion of wireless networks service and coverage into a future that promises both transparency and near ubiquity.

Indeed, we are now well past the early days of slow performance and lack of interoperability, and well into the proliferation of WLANs into a broad range of applications. We need to begin with an important baseline: although wireless networks use radio to communicate, they are still, in fact, networks. Anything that can be done on a traditional wired network can be done on a wireless network. And because modern IT infrastructures are networkcentric to the core, wireless networks fit into most installations with a minimum of fuss.

WiMAX without Wires

WiMAX wireless metropolitan area networks (MANs), based on the IEEE 802.16 family of standards, is a solution that can offer wireless broadband Internet access to residences and businesses at a relatively low cost. The standard supports shared transfer rates up to 75 Mbps from a single BS, which can offer broadband access without requiring a physical last mile connection from the end user to a service provider. Service delivery to end clients is likely to be roughly 300 kbps for residences and 2 Mbps for businesses.

The evolving 802.16 technology standard often referred to as broadband wireless or WiMAX potentially can deliver flexible, cost-effective fixed or portable wireless solutions enabling high-bandwidth services, with an array of multimedia features.

One of the promises of WiMAX is that it could offer the solution to what is sometimes called the last mile problem, referring to the expense and time needed to connect individual homes and offices to trunk lines for communications. WiMAX promises a wireless access range of up to 31 mi, compared to Wi-Fi's 300 ft and Bluetooth's 30 ft.

To appreciate what WiMAX brings to the table, we need to understand what additional features it provides over existing technologies. Existing BWA technologies that are closest to WiMAX with respect to service features are Wi-Fi, LMDS or current Multi-Channel Multi-Point Distribution Systems (MMDS) and third-generation (3G) mobile. Let us first examine these three closely.

Wi-Fi has risen to become one of the most popular forms of wireless local area networking thanks to its open standards, high speed, and ability to handle network interference. Still, Wi-Fi's popularity has exposed its primary limitation — range. It can only serve signals in a "hot spot" with a typical reach of about 1000 ft (300 m) outside or 328 ft (100 m) indoors, because of interference.

Wi-Fi Will Not Provide Ubiquitous Broadband.

The biggest problem with 3G or 4G is the usage cost, which is because of factors such as totally revamping the infrastructure and high license fees.

3G Will Not Provide Affordable Broadband.

WiMAX can satisfy a variety of access needs. Potential applications include extending broadband capabilities to bring them closer to subscribers, filling gaps in cable, DSL, and T1 services, Wi-Fi and cellular backhaul, providing "last 100 m" access from fiber-to-the-curb, and giving service providers another cost-effective option for supporting broadband services.

As WiMAX can support very-high-bandwidth solutions, in which large spectrum deployments (i.e., >10 MHz) are desired, it can leverage existing infrastructure, keeping costs down, while delivering the bandwidth needed to support a full range of high-value, multimedia services. Further, WiMAX can help service providers meet many of the challenges they face because of increasing customer demands without discarding their existing infrastructure investments, because it has the ability to seamlessly interoperate across various network types.

WiMAX can provide wide area coverage and quality-of-service capabilities for applications ranging from real-time delay-sensitive VoIP to real-time streaming video and non-real-time downloads, ensuring that subscribers get the performance they expect for all types of communications.

WiMAX, which is an IP-based wireless broadband technology, can be integrated into both wide-area 3G mobile and wireless and wireline networks, allowing it to seamlessly become part of an anytime, anywhere broadband access solution.

CNET Networks nominated WiMAX the "Most Promising Technology of the Year" in October 2003.

Making Broadband Personal

Broadband can be transformed from a limiting wired experience to an anytime, anywhere personal experience for use at home, office, or on the move. Plug-and-play portability can turn wireless broadband from a facilities-based solution into a personal solution with many potential vertical applications.

Key requirements for making broadband personal are the following:

- Broadband rates and reliability: competitive with DSL/cable
- Economics: competitive with the wired alternatives such as DSL/cable
- Zero install plug and play: no truck rolls/no technical installation
- Secured and private: no gate-crashing or spam boom
- Portability and mobility: deliver the freedom of wireless

Existing technologies are not sufficient to deliver on the promise of anytime, anywhere, as current technologies compromise on at least one of the following key attributes:

3G cellular is long range, mobile, reliable, plug and play, secure, private, and has manageable data rates, but data applications are too expensive, as much as ten times the cost of using similar wireline services.

Fixed wireless is long range, reliable, secure, private, low cost and has very high data rates, but it is neither mobile nor plug and play.

Wireless LAN is reliable, private, plug and play, low cost, nomadic, and has very high data rates, but its range is very short and it is not very well secured.

Mobile WiMAX or IEEE 802.16e is the right standard for personal broadband as it is a low-cost, high-performance long-range mobile solution for delivering secured broadband wireless data at high rate (Figure 7.6).

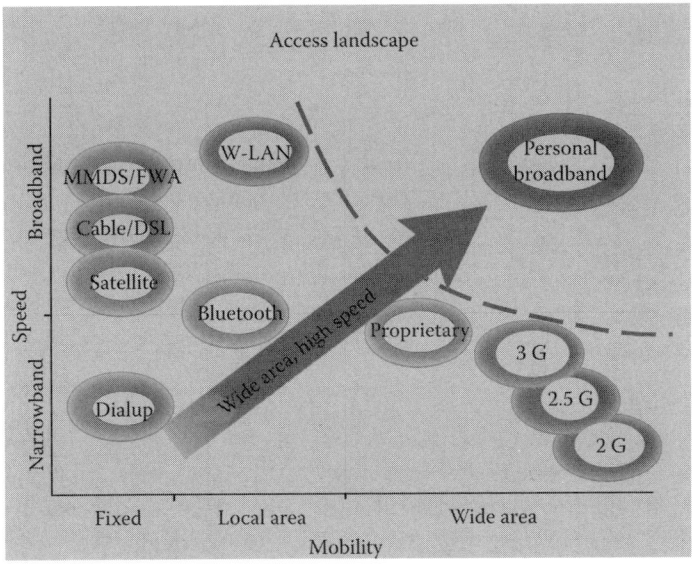

Figure 7.6 Access technologies for personal broadband.

One for Everyone

There are many technologies in various phases of evolution technically capable of providing BWA. In the first instance, these different BWA technologies look very similar; they also compete with each other. This is obvious as all of them provide broadband access wirelessly. In reality, the technologies are not competing head-on at all, although all of them can do voice and all can also do data at high speed, so in many situations they do compete. However, if carefully evaluated, they seem to be very different as well as synergic. Each of these technologies provides opportunities in different areas of application and deployment. It is a question of what the user needs, and hinges on three variables: distance, cost, and what services each solution is deployed for.

Although there is all-round euphoria in all these BWA camps, three technologies seems to be most appropriate today, in terms of cost, reliability, market acceptance, and performance. These are seemingly competing technologies with different standards vying to solve the BWA puzzle — Wi-Fi, WiMAX, and cellular technology (2.75G and higher).

Each technology has its own merits and demerits, as is the case with Wi-Fi, WiMAX, and fixed wireless.

Strengths
Wi-Fi
Convenience: Continuous, wireless connection to a corporate network or the Internet from a variety of sites (airports, hotels, restaurants, offices, hospitals, homes, etc.); improving worker connectivity and, therefore, productivity.

Compatibility: Connections to PCs, laptops, and PDAs with a wireless LAN card adhering to IEEE's 802.11b (or other) standard.

Interoperability: A nonproprietary, standardized solution.

Ad hoc mode: Direct communication between two compatible 802.11 devices without an access point (BS).

Installation speed and flexibility: Fast and easy to install, eliminating the need to cable the desktop.

Scalability: Modular configurations to suit changing density requirements.

WiMAX
Cost: By enabling standards-based products with fewer variants and larger volume production, it will drive the cost of equipment down.

Competition and choice: Having standardized equipment will also encourage competition, making it possible to buy from many sources.

Ease of deployment: Because of certified and standard equipments, it makes an excellent case for plug-and-play installation.

Reach: Can serve distances up to 25 to 30 km.

Spread: Can scale to support thousands of users with a single BS.

Cellular
Reach: Can serve large distances.

Installation speed and flexibility: Fast and easy to install, eliminating the need for digging, cabling, etc.

Ease of maintenance: Does not have components requiring high maintenance, very low fault rate, fast and easy turnaround due to low system complexity.

Weaknesses
Wi-Fi

Security: Opens the network to the public. Anyone with Wi-Fi compatibility has access to the network. A number of problems such as "war chalking" and "war driving" are new phenomena that are only recently becoming apparent. This is when hackers "drive" around and "chalk" the frequency of a Wi-Fi onto the ground for other hackers. Theoretically, anyone with an 802.11b/Wi-Fi client device can tap into your network via a nonsecure access point.

Cost: Low-volume chip production for client device makes cost of solution high, as cost depends heavily on the client device, which in turn depends on the chip.

Range: Short range of 200 m (can be enhanced using high-cost proprietary devices) for standardized solution.

WiMAX

Availability: Not yet widely available, encouraging numbers possible only by 2006.

Infrastructure: Requires additional backhaul to feed wireless network, BSs etc.

Spectrum: Uses both licensed and unlicensed bands.

Cellular

Cost: As it involves high level of IPR, requirement of high-cost installations, and hefty license fees, it is quite expensive.

Thus, no technology is a clear winner, but each supplements or complements another. If Wi-Fi is a strong contender for high-mobility indoor enterprise application, then WiMAX is just about perfect for multiple-site mass metropolitan applications, although cellular technology provides a more than appealing solution for high-speed mobile requirements.

In future, we are most likely to see hybrid wireless network technologies providing anytime and anywhere broadband access as no single technology can completely cater to this demand. Each wireless technology is designed to serve a specific usage segment (referred to as PANs, LANs, MANs, and WANs) based on many variables, including bandwidth, distance, power, user location, services offered, network ownership, and coverage. In a hybrid wireless network, a combination of two or more such technologies are deployed, leading to enhanced performance in terms of coverage, capacity, and throughput as compared to access networks based on a single technology.

There is a wealth of BWA technologies and associated developments. Optimized wireless access technologies exist for each usage segment and applications. High-frequency radio above 20 GHz aims at a large capacity to serve demanding users, whereas lower-frequency radio aims toward improved quality of service, and increased capacity and coverage, and 3G mobile provides excellent broadband access with high-speed mobility.

Wi-Fi, WiMAX, 3G, fixed wireless access (FWA) and UWB technologies each are necessary to form the global wireless infrastructure needed to deliver uninterrupted high-speed communication and seamless broadband Internet access worldwide.

Although Wi-Fi is ideal for isolated "islands" of connectivity, WiMAX and 3G are needed for long-distance wireless "canopies." Meanwhile, WiMAX and 3G are both required because their optimum platforms differ: WiMAX works best for computing platforms such as laptops, whereas 3G is best for mobile devices such as PDAs and cell phones. UWB offers very-short-range connectivity, perfect for the home entertainment environment or wireless USB. In short, each technology is important for different reasons.

It is not a case of one technology becoming universal, or of one technology replacing another. All the wireless networks will get built out for different usages, with some overlap at the edges. But most importantly, the technologies will coexist, creating more robust solutions that will enable many new and exciting possibilities.

The Way Ahead

How could the wireless communications competitive marketplace be characterized?

As any economic phenomenon, the wireless communications market is driven by economic forces. The key trend of the industry worldwide is the shift of value creation down the value chain to the customer. Companies that own customer relationships and actively exploit them by providing services and information will reap the biggest profits, and those that decide to stay in a more backbone-related area will very soon find themselves competing exclusively on price in a commodity-type business.

What are the reasons for this change?

The underlying factor is not difficult to guess — technological innovation.

3G and Beyond

It is worth briefly relating these data communications standards to the evolution of conventional cellular telephony as it adds data communications services. 802.16 starts from the premise of delivering broadband data to fixed points. For example, it generally assumes a reasonably wide channel allocation. To this it is adding mobility capabilities via 802.16e, which will allow it to support at least a nomadic model; i.e., one in which an end station does not move much while operating but may move around between sessions. It has also been adding support for narrow channels. Higher degrees of mobility will also require considerably more support for handing active connections off between BSs. On this infrastructure, one can then think about running VoIP to provide standard telephony service.

3G cellular systems start with the premise of delivering highly mobile voice services and increasing narrow- to mid-bandwidth data services. Their infrastructure is optimized for high mobility, including high-speed handoff. Data services are carried over a somewhat more complex technical structure designed for these needs.

For the immediate future, what infrastructure to deploy will be determined by previous investments in infrastructure (e.g., an existing 2G cellular system) or the specific needs (e.g., good data services to remote rural areas with little mobility needs). Looking forward, 4G systems are at this point primarily just a name A good working assumption, however, is that 4G systems will be a marriage of the best attributes of 3G cellular and packet-based wireless access systems (Figure 7.7).

There is still uncertainty about where such standards will be developed. Standards have been defined for data mainly in IEEE, whereas cellular standards have come from the ITU. A new group within the IEEE, 802.20, has begun to look at highly mobile systems from a datacentric perspective. At this point, it is too early to decide how 802.20, continued improvements to 802.16, and various possible 3G follow-on standards will relate to one another.

One other trend worth noting is toward end-user devices that can interact with multiple types of networks. Clearly, PCs are intelligent enough devices that they can easily support multiple radios or flexible radios, which will permit them to communicate using multiple standards

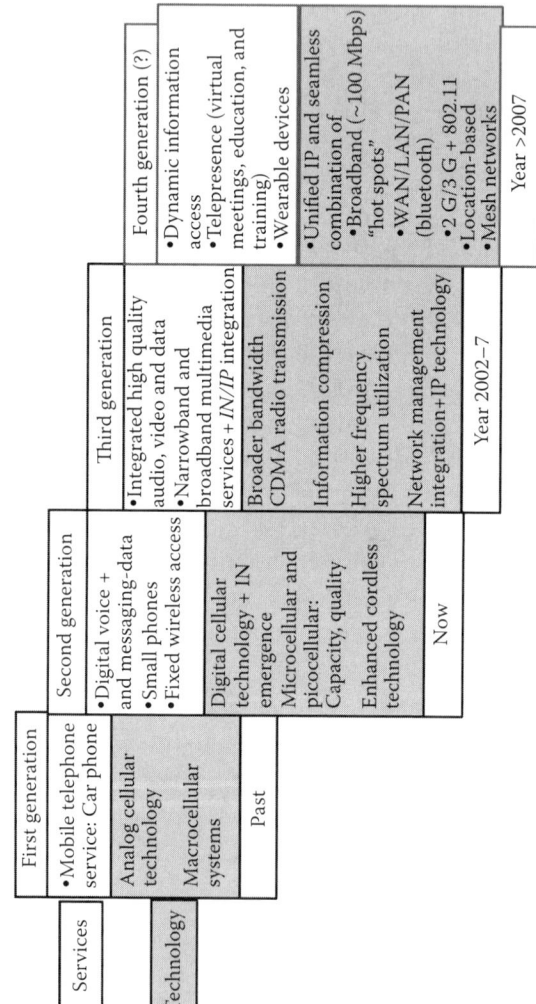

Figure 7.7   Network services and technology evolution to 4G.

choosing the best available network, be it 802.11, 802.16, or a wide area cellular technology such as GPRS. Smaller devices, including PDAs and voice handsets, will also evolve to support multiple networks. For example, a carrier may support a handset that uses VoIP over 802.11 if it finds itself in a place where such a WLAN is present, but switch to a standard cellular mode otherwise. Considerable technical work is still required to define the standards needed to make smooth transitions across such disparate networks work well.

The following text summarizes information about 802.16 and WiMAX:

The Power of 802.16
Designed from the ground up for WANs
True broadband systems for multi-Mbps services to users
High capacity and ease of deployment
Carrier-class features and reliability
Scalability and guaranteed service levels

Convergence (Figure 7.8)
Bringing IEEE 802.16 and ETSI HIPERMAN together
Voice, video, and data in a unified IP network

Interoperability
Enabler for the high-volume worldwide market

WiMAX Official Timeline (Figure 7.9)
IEEE 802.16 Rev 2004 has been published.
Baseline for the WiMAX interoperability profiles and test cases.
Replaces 802.16a (but with no compatibility).
Next revision will be 802.16e (completion planned for Q1 2005).
WiMAX interoperability profiles and certification is in progress.
Completion date for certification tests is late as was slotted for 2004.
Will lead to first "plug-fests" in Q1 2005.
Second and third plug-fests to complete early Q2 2005.
First WiMAX-certified shipping systems are expected in middle or late 2005.
Low-cost WiMAX CPEs are not expected before 2006.

A Path to Mobility
IEEE 802.16 defines an evolution path toward a mobile system.
802.16 Rev D for fixed applications.
802.16e for mobile and portable applications.
Draft of the 802.16e standard is now moving to working group ballot.
PHY layer should remain very similar to Rev D.
802.16e BS will support Rev D CPEs.
Completion of the standard is expected in the first half of 2005.
This could lead to first systems in the 2007 time frame.

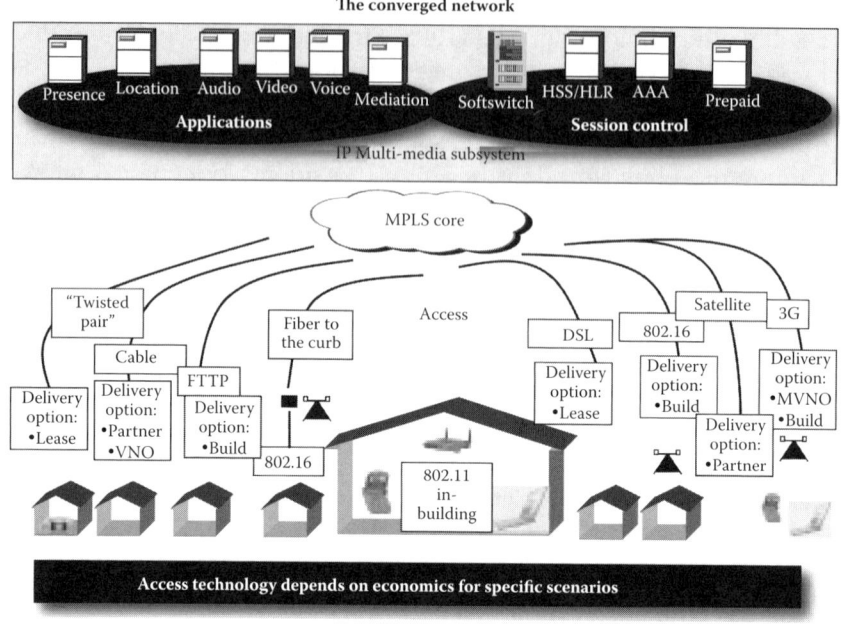

Figure 7.8   Convergence provides economic access technology options.

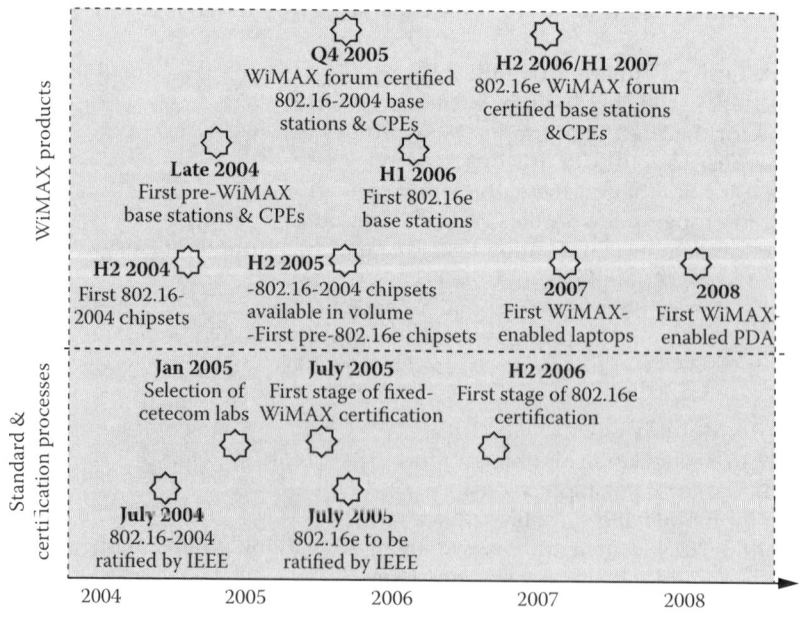

Figure 7.9   WiMAX timeline.

Chapter 8

Understanding the Technology

Less than one out of five people of the developed world and an even smaller, minuscule percentage of people across the world have broadband access today. Existing technologies such as DSL, cable, and fixed wireless are plagued by expensive installs, problems with loop lengths, upstream upgrade issues, line-of-sight restrictions, and poor scalability.

WiMAX is the next step on the road to a broadband as well as a wireless world, extending broadband wireless access to new locations and over longer distances, as well as significantly reducing the cost of bringing broadband to new areas. WiMAX technology offers greater range and bandwidth than the other available or soon-to-be available broadband wireless access technologies such as wireless fidelity (Wi-Fi) and Ultrawideband (UWB) family of standards and provides a wireless alternative to wired backhaul and last mile deployments that use Data Over Cable Service Interface Specification (DOCSIS) cable modems, Digital Subscriber Line technologies (xDSL), T-carrier and E-carrier (Tx/Ex) systems, and Optical Carrier Level (OC-x) technologies.

WiMAX technology can reach a theoretical 30 mi coverage radius and achieve data rates up to 75 Mbps, although at extremely long range, throughput is closer to the 1.5 Mbps performance of typical broadband services (equivalent to a T1 line), so service providers are likely to provision rates based on a tiered pricing approach, similar to that used for wired broadband services.

The overall concept of metropolitan area wireless networking, as envisioned with 802.16, begins with what is called *fixed wireless*. Here, a backbone of base stations is connected to a public network, and each base station supports hundreds of fixed subscriber stations, which can be both public Wi-Fi hot spots and firewalled enterprise networks. Later in the development cycle, with 802.16e, WiMAX is expected to support mobile wireless technology — that is, wireless transmissions directly to mobile end users. This will be similar in function to the General Packet Radio Service and the one times radio transmission technology (1×RTT) offered by phone companies.

New enterprises as well as individuals are increasingly adopting broadband, whereas those already using broadband are becoming dependent on it and are demanding better services with added benefits. To support this unprecedented new demand, wireless broadband has emerged as a viable solution. WiMAX, because of its inherent features, holds great promise for the future of broadband communications.

There has been a lot of hype about WiMAX and the impact that this standards-based wireless broadband network technology will have on the broadband access market. All this hype has generated tremendous expectations, and the industry has responded with exceptional aggression and commitment toward taking broadband to the next level with WiMAX.

How WiMAX Works

Let us take a quick glance at the working of a basic WiMAX system.

A WiMAX base station is connected to public networks using optical fiber, cable, microwave link, or any other high-speed point-to-point (P-P) connectivity, referred as a *backhaul*. In few cases such as mesh networks, point-to-multi-point (P-MP) connectivity is also used as a backhaul. Ideally, WiMAX should use point-to-point antennas as a backhaul to connect aggregate subscriber sites to each other and to base stations across long distances.

A base station serves subscriber stations (also called customer premise equipment [CPE] for obvious reasons) using non-line-of-sight (NLOS) or line-of-sight (LOS) point-to-multi-point connectivity, and this connection is referred to as the *last mile*. Ideally, WiMAX should use

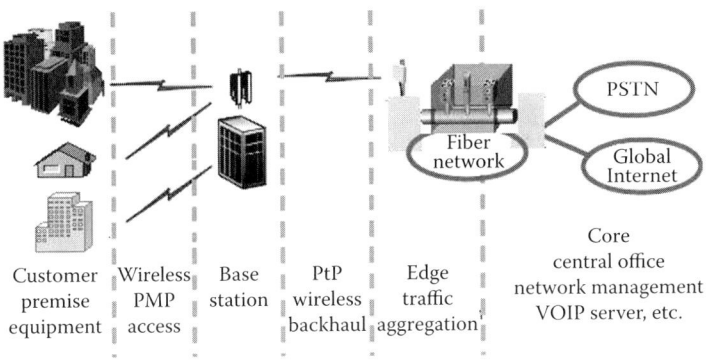

Customer premise equipment | Wireless PMP access | Base station | PtP wireless backhaul | Edge traffic aggregation | Core central office network management VOIP server, etc.

Figure 8.1   WiMAX network.

NLOS point-to-multi-point antennas to connect residential or business subscribers to the base station (Figure 8.1).

A subscriber station typically serves a building (business or residence) using wired or wireless LAN.

Designed to Succeed

WiMAX has been designed to address challenges associated with traditional wired and wireless access deployments. Although the backhaul connects the system to the core network, it is not an integrated part of WiMAX system as such.

Typically, a WiMAX system consists of two parts' a WiMAX base station and a WiMAX receiver (also referred as CPE).

WiMAX Base Station

A WiMAX base station consists of indoor electronics and a WiMAX tower. Typically, a base station can cover up to 6 mi radius (theoretically, a base station can cover up to 50 km radius or 30 mi, but practical considerations limit it to about 10 km or 6 mi). Any wireless node within the coverage area would be able to access the Internet (Figure 8.2).

The WiMAX base stations would use the media access control layer defined in the standard (a common interface that makes the networks interoperable) and would allocate uplink and downlink bandwidth to subscribers according to their needs, on an essentially real-time basis.

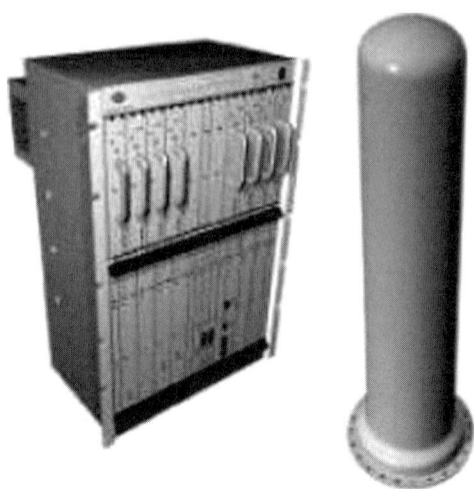

Figure 8.2 WiMAX base station.

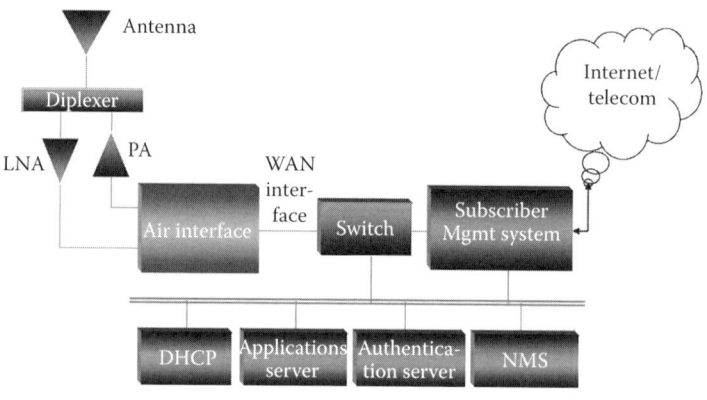

Figure 8.3 Base station architecture.

WiMAX Receiver

A WiMAX receiver, which is also referred as CPE, may have a separate antenna (i.e., receiver electronics and antenna are separate modules) or could be a stand-alone box or a PCMCIA card that sits in a laptop or computer. Access to a WiMAX base station is similar to accessing a wireless access point (AP) in a Wi-Fi network, but the coverage is more (Figure 8.4).

Figure 8.4   Indoor WiMAX CPEs.

So far one of the biggest deterrents to the widespread acceptance of broadband wireless access (BWA) has been the cost of CPE. This is not only the cost of the CPE itself, but also that of installation. Historically, proprietary BWA systems have been predominantly LOS, requiring highly skilled labor and a truck role to install and "turn up" a customer. The concept of a self-installed CPE has been the Holy Grail for BWA from the beginning. With the advent of WiMAX, this issue seems to be getting resolved.

Backhaul

Backhaul refers both to the connection from the AP back to the provider and to the connection from the provider to the core network. A backhaul can deploy any technology and media provided it connects the system to the backbone. In most of the WiMAX deployment scenarios, it is also possible to connect several base stations with one another by use of high-speed backhaul microwave links. This would also allow for roaming by a WiMAX subscriber from one base station coverage area to another, similar to roaming enabled by cellular phone companies.

Flavors of WiMAX

WiMAX can provide two flavors of wireless services, depending on the frequency range of operation. These frequency ranges are 10 to 66 GHz and 2 to 11 GHz. The microwave frequencies below 10 GHz

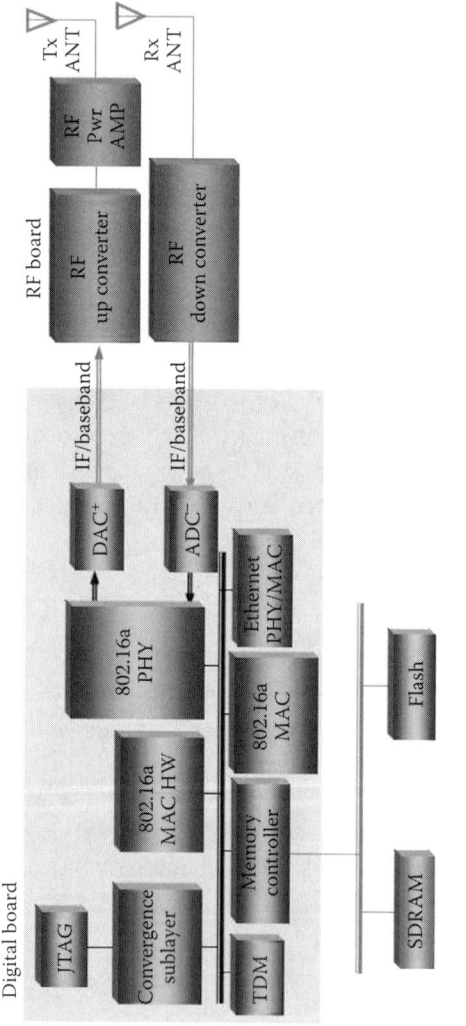

Figure 8.5 WiMAX CPE architecture.

are referred to as *centimeter bands*. Above 10 GHz, they are known as *millimeter bands*.

Millimeter bands have much wider allocated channel bandwidths to accommodate the larger data capacities that are suitable for high-data-rate, LOS backhauling applications. Centimeter bands are best for multi-point, near-line-of-sight, tributary, and last mile distribution.

Line-of-Sight

The original 802.16 standard operates in the 10 to 66 GHz frequency band and requires LOS towers. The LOS access service employs a dish antenna that points straight at the WiMAX tower from a rooftop or pole. The LOS connection is stronger and more stable, so it is able to send a lot of data with fewer errors. LOS transmissions use higher frequencies, with ranges reaching a possible 66 GHz. At higher frequencies, there is less interference and more bandwidth.

Through the stronger LOS antennas, the WiMAX transmitting station would send data to WiMAX-enabled computers or routers set up within the transmitter's 30 mi radius (3600 sq mi or 9300 sq km of coverage). This is what allows WiMAX to achieve its maximum range.

Non-Line-of-Sight

The 802.16a extension, ratified in January 2003, uses a lower frequency of 2 to 11 GHz, enabling NLOS connections. This was a major breakthrough in wireless broadband access because LOS between transmission point and the receiving antenna is not necessary. With 802.16a, more customers can be connected to a single tower, substantially reducing service costs.

The NLOS access service is very similar to Wi-Fi, in which a small antenna on a computer connects to the tower. Lower-frequency transmissions are not as easily disrupted by physical obstructions as the high-frequency transmissions, and they are better able to diffract, or bend, around obstacles. Based on this principle, WiMAX uses a lower frequency range of 2 GHz to 11 GHz (similar to Wi-Fi) in this mode.

NLOS-style access will be limited to a radius between 4 to 6 mi (perhaps 25 sq mi or 65 sq km of coverage, which is similar in range to a cell phone zone).

The centimeter spectrum contains both tributary and last mile potential. IEEE 802.16-2004 supports fixed-NLOS BWA to supplant or supplement DSL and cable access for last mile service.

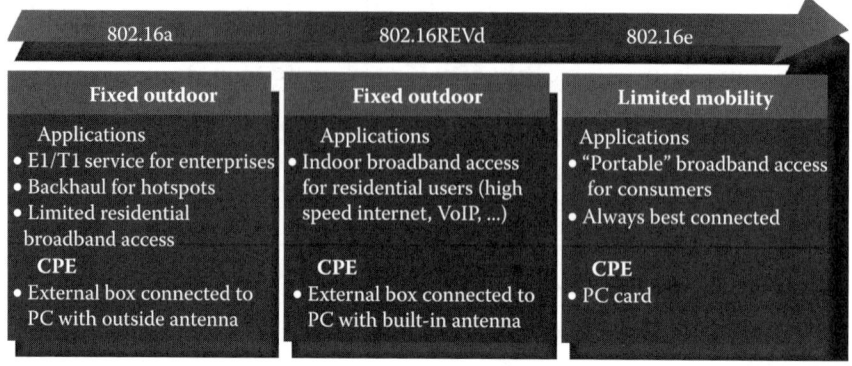

Figure 8.6   WiMAX flavors.

Types of WiMAX

The WiMAX family of standards addresses two types of usage models: a fixed-usage model (IEEE 802.16-2004) and a portable usage model (802.16 REV E, scheduled for ratification in current year).

Before we discuss more about these distinct types of WiMAX, it is important to understand and appreciate key differences between the mobile, nomadic, and fixed wireless access systems. The basic feature that differentiates these system is the ground speed at which the systems are designed to operate. Based on mobility, wireless access can be divided into four classes: stationary (0 km/hr), pedestrian (up to 10 km/hr), and vehicular (subclassified as "typical" up to 100 km/hr and "high speed" up to 500 km/hr).

A mobile wireless access system is one that can address the vehicular class, whereas the fixed serves the stationary and pedestrian classes. This raises a question about the nomadic wireless access system, which is referred to as a system that works as a fixed wireless access system but can change its location. An example is a WiMAX subscriber operating from one location, i.e., the office during daytime, and moving to another location, i.e., the residence in the evening. If the wireless access system works at both the locations, it can be referred to as *nomadic*.

Fixed

Service and consumer usage of 802.16 for fixed access is expected to mirror that of fixed wireline service, with many of the standards-based requirements being confined to the air interface. Because communication takes place via wireless links from CPE to a remote NLOS base station, requirements for link security are greater than those needed for wireline service. The security mechanisms within the IEEE 802.16 standards are adequate for fixed access service.

An additional challenge for the fixed-access air interface is the need to establish high-performance radio links capable of data rates comparable to wired broadband service, using equipment that can be self-installed indoors by users, as is the case for DSL and cable modems. IEEE 802.16 standards provide advanced physical (PHY) layer techniques to achieve link margins capable of supporting high throughput in NLOS environments.

Portable or Mobile

The 802.16a extension, ratified in January 2003, uses a lower frequency of 2 to 11 GHz, enabling NLOS connections. The latest 802.16e task group is capitalizing on the new capabilities this provides by working on developing a specification to enable mobile 802.16 clients. These clients will be able to hand off between 802.16 base stations, enabling users to roam between service areas.

There can be two cases of portability: full mobility or limited mobility. The simplest case of portable service (referred to as *nomadicity*) involves a user transporting an 802.16 modem to a different location. Provided this visited location is served by wireless broadband service, in this scenario the user reauthenticates and manually reestablishes new IP connections and is afforded broadband service at the visited location.

In the fully mobile scenario, user expectations for connectivity are comparable to facilities available in third-generation (3G) voice/data systems. Users may move around while engaged in a broadband data access or multimedia streaming session. Mobile wireless access systems need to be robust against rapid channel variation to support vehicular speeds.

There are significant implications of mobility on the IP layer owing to the need to maintain routability of the host IP address to preserve in-flight packets during IP handoff. This may require authentication, and handoffs for uplink and downlink IP packets and MAC frames. The need to support low latency and low-packet-loss handovers of data streams as users transition from one base station to another is clearly a challenging task. For mobile data services, users will not

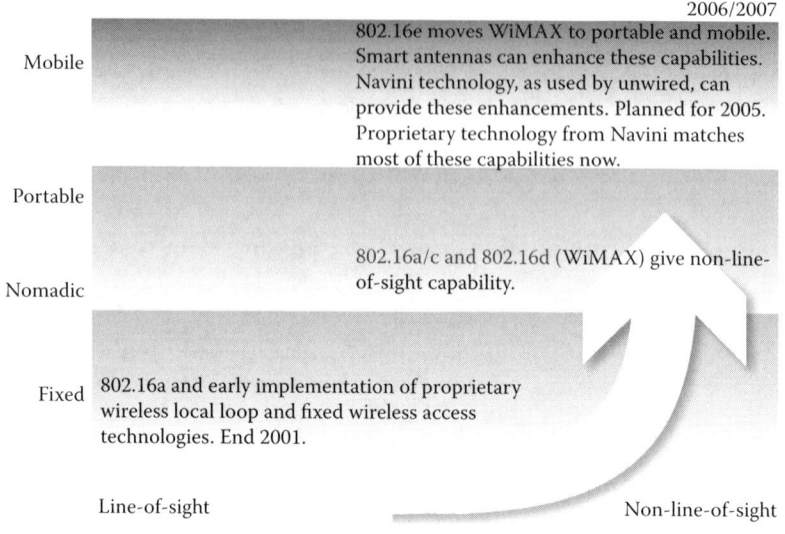

Figure 8.7 WiMAX types.

easily adapt their service expectations because of environmental limitations that are technically challenging but not directly relevant to the mode of user (such as being stationary or moving). For these reasons, the network and air interface must be designed to anticipate these user expectations and deliver accordingly.

IEEE 802.16e will add mobility and portability to applications such as notebooks and PDAs. Both licensed and unlicensed spectrums will be utilized in these deployments. 802.16e is tentatively scheduled to be approved in the second half of this year.

Evolution of WiMAX

The search for a broadband wireless delivery system began several years ago with two actions by Congress and the Federal Communication Commission (FCC): the NII (National Information Infrastructure) bands in the 5 to 6 GHz range and the creation of Local Multi-Point Distribution Service (LMDS) around 30 GHz. These new services were intended to be used for wireless delivery of Internet access and other broadband services, as both an alternative to DSL and cable modem, and as a new service to areas where a wired infrastructure was difficult to install or was not economically feasible.

LMDS got the greatest attention from investors as it promised to offer broadband Internet along with various exciting services including entertainment services in competition with cable television (CATV) systems. The reason for high expectations was the bandwidth allocated to LMDS, which was sufficient for providing multimedia services. Companies that started working on LMDS got plenty of publicity about their efforts to create a new wireless broadband marketplace. Their failure to achieve the stated goals was both a discouragement to the investment community and a reason for the engineering community to rethink how the technology should work.

As the story of these early ventures was playing out, a new standards effort was begun by an IEEE committee, IEEE 802.16, which soon became known as WirelessMAN (metropolitan area network). The 802.16 committee began exploring not only LMDS but also potential licensed and unlicensed services from 2 to 66 GHz. Later, its scope was increased further, with mobile services added to the original fixed-service approach. WirelessMAN was conceived to provide a more usable approach to wireless broadband, a better fit to the manner in which customers are expected to use it.

At the same time, parallel with the IEEE 802.16 standards activity, private organizations involved in the development of wireless broadband formed an industry consortium, the WiMAX Forum, which was established in 2001. This effort is similar to the Wi-Fi consortium that coordinates component and equipment development for IEEE 802.11 wireless LAN.

The WiMAX Forum is complementary to the standards activities. The IEEE committees simply create the operating standards. The WiMAX Forum has the purpose of promoting voluntary coordination among companies to ensure that company equipment will interoperate. For a market to develop, equipment operation must be reliable, and an independent means of ensuring interoperability is essential. Equipment that has been verified through a WiMAX Forum laboratory will be designated "WiMAX Forum Certified™" to provide assurance to the consumer.

The Cutting Edge

WiMAX is an integrated suite of many innovative and advance techniques covering diverse areas such as modulation, antenna diversity, interference, etc. Some of the key developments having a direct bearing

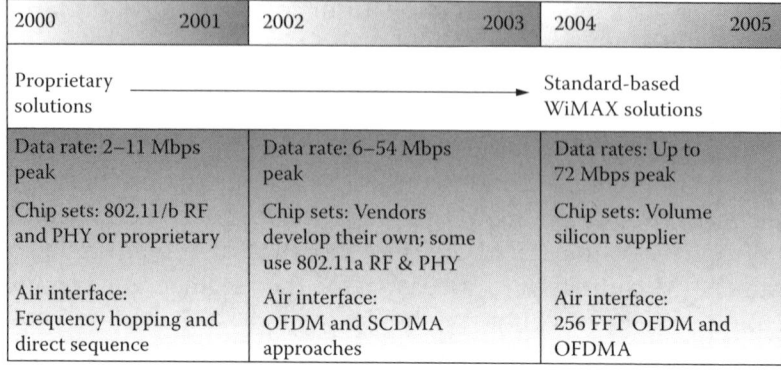

2000	2001	2002	2003	2004	2005
Proprietary solutions				Standard-based WiMAX solutions	
Data rate: 2–11 Mbps peak		Data rate: 6–54 Mbps peak		Data rates: Up to 72 Mbps peak	
Chip sets: 802.11/b RF and PHY or proprietary		Chip sets: Vendors develop their own; some use 802.11a RF & PHY		Chip sets: Volume silicon supplier	
Air interface: Frequency hopping and direct sequence		Air interface: OFDM and SCDMA approaches		Air interface: 256 FFT OFDM and OFDMA	

Figure 8.8 WiMAX evolution.

on commercial acceptance of system functionality are described in the following subsections.

Dynamic Burst Mode TDMA MAC: Provides High Efficiency, Bandwidth on Demand, and Scalability

802.16 is optimized to deliver high, bursty data rates to subscriber stations. This means that IEEE 802.16 is uniquely positioned to extend broadband wireless beyond the limits of today's systems, both in distance and in the ability to support applications.

QoS: A Powerful WiMAX Advantage

Several features of the WiMAX protocol ensure robust quality-of-service (QoS) protection for services such as streaming audio and video. As with any other type of network, users have to share the data capacity of a WiMAX network, but WiMAX's QoS features allow service providers to manage the traffic based on each subscriber's service agreements on a link-by-link basis. Service providers can therefore charge a premium for guaranteed audio/video QoS, beyond the average data rate of a subscriber's link.

Improved User Connectivity

WiMAX keeps more users connected by virtue of its flexible channel widths and adaptive modulation. Because it uses channels narrower than the fixed 20 MHz channels used in 802.11, the 802.16-2004 standards can serve lower-data-rate subscribers without wasting bandwidth. When subscribers encounter noisy conditions or low signal strength, the adaptive modulation scheme keeps them connected when they might otherwise be dropped.

Link Adaptation: Provides High Reliability

WiMAX provides adaptive modulation and coding — subscriber by subscriber, burst by burst, and uplink and downlink. Transmission adaptation with the help of modulation depending on channel conditions provides high reliability to the system. Further, this feature imparts differential service provision, making the system economically more appealing to operators.

Intelligent Bandwidth Allocation: Provides Guaranteed Service Levels

Terminals have a variety of options available to them for requesting bandwidth, depending on the QoS and traffic parameters of their services. The option of bandwidth on demand (frame by frame) by reallocation of frequency band makes WiMAX flexible as well as efficient.

NLOS Support: Provides Wider Market and Lower Costs

WiMAX solves or mitigates the problems resulting from NLOS conditions by using multiple frequency allocation support from 2 to 11 GHz, orthogonal frequency division multiplexing (OFDM) and orthogonal frequency division multiple access (OFDMA) for NLOS applications (licensed and license-exempt spectrum), subchannelization, directional antennas, transmit and receive diversity, adaptive modulation, error correction techniques, and power control.

Highly Efficient Spectrum Utilization

In WiMAX, the MAC is designed for efficient use of spectrum and incorporates techniques for efficient frequency reuse, deriving a more efficient spectrum usage of the access system.

Secured Data Exchange

WiMAX proposes the full range of security features to ensure secured data exchange: terminal authentication by exchanging certificates to prevent rogue devices, user authentication using the Extensible Authentication Protocol (EAP), data encryption using the Data Encryption Standard (DES) or Advanced Encryption Standard (AES), both of which are much more robust than the Wireless Equivalent Privacy (WEP) standard initially used by WLAN. Furthermore, each service is encrypted with its own security association and private keys.

IEEE 802.16 Standards

Since its early days IEEE 802.16 standards have seen many changes. Even today innovation continues, and the standards evolve with every passing day and new technological advance. We now describe these standards and the subsequent changes in detail.

WiMAX is an international undertaking, a global wireless access technology that addresses interoperability across products based on the IEEE 802.16 standard. IEEE 802.16 is an emerging global broadband wireless access standard capable of delivering multiple megabits of shared data throughput supporting fixed, portable, and mobile operation.

As with other IEEE 802.XX standards, IEEE 802.16 is actually a family of standards, some completed and some still in progress (IEEE has accelerated its standards-making in response to market demand). The family of IEEE 802.16 standards offers a great deal of design flexibility, including support for licensed and license-exempt frequency bands, channel widths ranging from 1.25 to 20 MHz, QoS establishment on a per-connection basis, strong security primitives, multicast support, and low latency/low-packet-loss handovers.

This family also has an associated industry group, the WiMAX Forum.

Overview

The original 802.16 standard defines a MAC suitable for an access system based at a central base station serving multiple users scattered over a relatively large area whose radius can be many miles. This version of the standard also defines a particular PHY layer that is suited for use in bands between 10 and 66 GHz. The standard is optimized

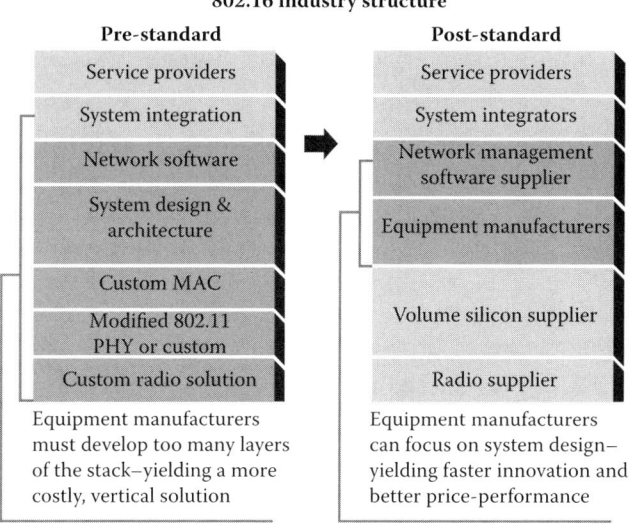

Figure 8.9   Pre- and post-standards WiMAX industry.

for providing an access service to, for example, an entire building in which multiple systems can all be attached to a single transceiver.

It is helpful to think of 802.16 as an alternative to a wire or fiber for delivering an Internet connection to a site. Unlike 802.11, this standard does not narrowly define the frequency bands that should be used nor does it limit the channel width. It can therefore accommodate any band, licensed or unlicensed, that may be available to an operator. For example, with a 20 MHz channel available and a strong signal between the base station and remote site, the capacity can be as high as 96 Mbps, whereas a 28 MHz channel in a similar situation could achieve over 130 Mbps. On the other hand, remote sites at the limit of reception of a base station might only be able to achieve a 32 Mbps capacity. Of course, as a multi-point system expected to serve many customers, these capacities must be shared across all the served users.

Family of IEEE 802.16 Standards

The Institute of Electrical and Electronics Engineers Standards Association (IEEE-SA) sought to make broadband wireless access more widely available by developing IEEE Standard 802.16, which specifies the WirelessMAN Air Interface for wireless metropolitan area networks. The standard, which was published on 8 April, 2002, was created in a two-year, open-consensus process by hundreds of engineers from the world's leading operators and vendors. The term "802.16" is generically used to cover all the 802.16 revisions that service providers can choose from.

Line-of-Sight

One key issue for wireless access systems such as 802.16 is whether they require LOS between the receiver and the base station or whether near LOS or NLOS is sufficient. Ideally, LOS would not be required, but the reality is that radio waves are always attenuated when passing through obstacles so that NLOS performance will always be poorer than LOS performance. For good performance, the best design is an end station that mounts at least the antenna on the outside of the building facing generally in the direction of the base station. Although possibly complicating installation slightly, this will ensure the best range and performance for an 802.16 deployment.

For clarity on LOS issues, standards work has been divided into two frequency ranges, 2 to 11 GHz and 10 to 66 GHz. The reason for the division is the nature of signal propagation. Above 10 GHz, signals travel in a strictly LOS manner. The transmitter and receiver must quite literally see each other. Precipitation and vegetation create significant attenuation. Below about 10 GHz, transmission paths can be maintained with some deviation from LOS. Refraction and diffraction can bend a signal around corners, and the penetration of buildings is better than at higher frequencies.

The following list provides a brief overview of key differences:

- 802.16: 10 to 66 GHz, LOS.
- 802.16a: 2 to 11 GHz, NLOS. Standard finalized in January 2003.

802.16a is optimized for operation at frequencies between 2 and 11 GHz. It is also more flexible in channel width choices, including channels as narrow as 1.75 MHz to allow it to be used where only small allocations are available. This version is attracting considerable commercial attention now because this range covers a number of popular bands found around the world.

The unlicensed bands described earlier fall into this range. 802.16a systems are long-range systems and therefore at first look do not appear attractive for use in unlicensed bands, where interference among competing operators could become a problem. However, particularly in rural and developing markets, it is likely that there will be sufficient unlicensed spectrum and little enough competition for it that operators may find its use quite reasonable.

There are also commonly available bands at 2.5 GHz and 3.4 GHz in various countries that are licensable for use with an access data service. 802.16a looks to be a good choice for these systems as well. An operator licensed for exclusive use of part of one of these bands could offer broadband wireless access in more densely populated urban or suburban areas without interference concerns.

■ 802.16c: 10 to 66 GHz, NLOS.
   The purpose of 802.16c is to develop 10 to 66 GHz system profiles to aid interoperability. Specifications Standard was published in January 2003.
   The IEEE 802.16 Working Group develops standards that address two types of usage models: a fixed usage model (IEEE 802.16-2004) and a portable usage model (802.16 REV E, scheduled for ratification in 2005).
■ 802.16d: based on 802.16 and 802.16a with some improvements.
   802.16d, also known as 802.16-2004, incorporates and makes obsolete the 802.16 and 802.16a standards. It also supports sub-11-GHz spectrums. This standard was finalized on June 24, 2004. Both time division duplex (TDD) and frequency division duplex (FDD) are transmission options available within 802.16d.

Portability

The nomadic standard (802.16-2004) was published in July 2004 to consolidate previously published base standards and amendments.

Mobility

Mobile broadband wireless access has a significantly distinct identity from fixed, nomadic, and portable wireless; that is why a separate working group within IEEE 802 was needed to address mobile wireless.

- 802.16e: Capability to provide mobility/portability.

  The IEEE 802.16e extension adds support for mobile subscriber stations. 802.16e provides "hooks" for mobility/portability at the MAC and PHY layers. It does not address mobility at higher layers. The mobile standard (802.16e) has reached a final draft, incorporating scalable signal modulation modes (SOFDMA) for the mobility standard. It is expected to be finalized and published in 2005. These systems will most likely require new hardware components on the subscriber modules and, potentially, some hardware changes in the base stations.

- 802.16f: Improve the coverage using mesh networking (subject to approval as present status is of an ad hoc committee.)

  A newly formed group within 802.16, the Mesh Ad Hoc committee, is investigating ways to further improve the coverage of base stations. Mesh networking allows data to hop from point to point, circumventing obstacles such as hills. Only a small amount of meshing is required for a large improvement in the coverage of a single base station. If this group's proposal is accepted, it will become Task Force F and develop an 802.16f standard.

- 802.16g: Capability to support mobility at higher layers and across backhaul. The standardization date has not been determined.

  It is too early to determine the changes required when this phase of the standard becomes available. The standard presently specifies the use of either OFDM or OFDMA. OFDMA is based on OFDM and combines time division and frequency division multiple access techniques for more efficient spectrum utilization.

Importance of Standards

IEEE 802.16 standards as discussed at length earlier are very important, and it is critical to realize this. Further details regarding the role played by standards are given in the following text.

Standards enable the best technical ideas from academic and industry developers to be combined and amplified. The process of creating the standard subjects the proposed ideas to broad review and generally results in considerable conceptual and technical improvements to any individual proposals.

Standards play a vital role in driving costs much lower and much sooner. This is because the existence of a standard creates a common market to which competing companies sell. Furthermore, the size of that common market is larger than any submarket based on proprietary approaches would be. In the high-technology world, costs are strongly a function of volumes so the creation of a single high-volume market leads to much lower costs than a fragmented market. Standards also permit higher degrees of integration of equipment, which also lowers costs.

Standards are critical for allowing users to move around in their countries and around the world with their equipment working properly. Common standards mean that the equipment produced in one follower country is compatible to the systems of all other countries following those standards.

Last but not least, standards give a universal technology platform to all irrespective of their company, country, or community, on which they can build. Standards automatically ensure a level playing field for all.

Technology Description

The 802.16x family of wireless networking standards defines a set of solutions for metropolitan area networks (MANs) with a range of up to 30 mi. Products based on 802.16a will provide broadband connectivity to individual workstations or to LANs (wired or wireless) within the 30 mi range of the 802.16 base stations. 802.16a-based networks do not have an LOS requirement. It is highly likely that some 802.16x products will have the capability to be chained together to extend the 30 mi range; there are technical issues relating to signal overlap and interference that will need to be worked out, but this should be a feasible scenario.

The bandwidth available at a base station will depend on the specific standard implemented. For example, 802.16a products are expected to have bandwidth capacities up to 75 Mbps. The bandwidth would be split among the receiving stations, so a 70 Mbps base station serving 100 recipients would be able to provide a theoretical maximum of 700 kbps per recipient.

As spectrum-governing authorities will determine the usable spectrum for various services and decide which portions of the spectrum will serve which specific segments, variations in the spectrum allocation will impact radio-frequency (RF) interface and deployment. Although

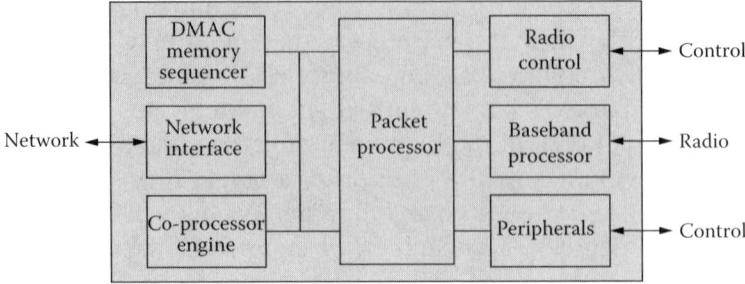

Figure 8.10   WiMAX system-on-a-chip (SoC).

a common RF ground exists, there is significant diversity in spectrum allocation and regulation. This diversity results in the demand for RF-diverse base stations and subscriber stations.

The WiMAX (IEEE 802.16) standard defines profiles for the MAC and PHY layers. The MAC layer packs and unpacks raw data, whereas the PHY layer handles the air interface and modulation schemes based on subscriber needs and RF link quality. The standard also allows system vendors to customize their products to meet specific requirements.

The interface between the RF front end and the system-on-a-chip (SoC) incorporates control signals for transmit and receive operations and housekeeping. It also houses I/Q signals to interface with analog-to-digital and digital-to-analog data converters. The receive data that is delivered by the demodulator circuit to the SoC should be differential "I" and "Q" signals. Attenuators can be employed on the receive side to handle calibration and gain control. They will ensure maximum bit usage and conversion efficiency.

The 802.16a standard also includes QOS features that are designed to enable voice and video transmission over wireless connections. The sophisticated MAC architecture can simultaneously support real-time multimedia applications requiring advanced QoS such as VoIP, streaming video, and online gaming.

Base Stations

WiMAX base stations can range from units that support only a few subscriber stations to elaborate equipment that supports thousands of subscriber stations and provides many carrier-class features. Whatever number of subscriber stations a base station supports, the latter must manage a variety of functions that are not required in subscriber equipment. Some base stations must support sophisticated antenna capabilities and implement efficient frequency reuse.

As a result, WiMAX base stations will have many different configurations. They will likely range from simple stand-alone units that support a few users to redundant, rack-mounted systems and server blades that operate alongside wireline networking equipment. On the hardware side, this equipment will typically use off-the-shelf microprocessors and discrete RF components.

Power Control

In any WiMAX network, power levels and control for both transmit and receive are important for system efficiency. To ensure successful communication, the levels must be actively managed. Power-control algorithms are used to improve the overall performance of the system. It is implemented by the base station sending power-control information to each of the CPEs to regulate the transmitted power level so that the level received at the base station is at a predetermined level.

Power control reduces the overall power consumption of the CPE and the potential interference with other colocated base stations. For LOS the transmit power of the CPE is approximately proportional to its distance from the base station, and for NLOS it is also heavily dependant on the clearance and obstructions. In a dynamical changing fading environment, this predetermined performance level means that the CPE only transmits enough power to meet this requirement. The converse would be basing the CPE transmit level on worst-case conditions.

Power levels are dynamically adjusted on a per-subscriber basis, depending on the profile and distance from the base station. For the base station transmitter, the actual transmitted power will depend on the subscriber distance, propagation characteristics, channel bandwidth, and modulation scheme (BPSK, QPSK, 16QAM, or 64QAM). The least data-efficient method is BPSK. Because it is employed where the subscriber station is farthest from the base, BPSK requires additional transmit power. 64QAM offers high data efficiency, which is best when the subscriber station is closer to the base station.

CPE or Subscriber System

Depending on the end-user needs, WiMAX provisioned for three different types of CPEs:

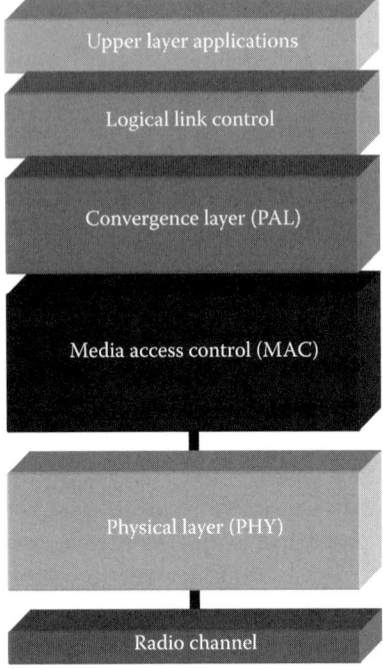

Figure 8.11 WiMAX protocol stacks reference model.

- ■ A modem attached to an external rooftop antenna
- ■ A modem with an indoor antenna
- ■ Integrated antenna, because as further integration into silicon by major chip suppliers takes hold, CPEs can be integrated into laptops, phones, and other devices.

A generic WiMAX subscriber system includes a control processor, MAC unit, base band processor (BBP), and analog RF front end. The front end places 802.16X in a specific licensed or unlicensed band.

IEEE 802.16 PHY Layer

The PHY layer has two complementary functions. One is to process data for transmission, where the output is a baseband I/Q signal or a 10 MHz IF signal. The process is reversed for the second function of receiving data, where the input is a baseband I/Q signal or a 10 MHz IF signal. For data reception, the PHY layer implements proprietary

Table 8.1 IEEE 802.16a PHY Layer Features: Benefits

Features	Benefits
256-point FFT OFDM waveform	Built-in support for addressing multi-path in outdoor LOS and NLOS environments
Adaptive modulation and variable error correction encoding per RF burst	Ensures a robust RF link while maximizing the number of bits/ second for each subscriber unit
TDD and FDD duplexing support	Addresses varying worldwide regulations where one or both may be allowed
Flexible channel sizes (e.g., 3.5 MHz, 5 MHz, 10 MHz, etc.)	Provides the flexibility necessary to operate in many different frequency bands with varying channel requirements around the world
Designed to support smart antenna systems	Smart antennas are fast becoming more affordable, and as these costs come down their ability to suppress interference and increase system gain will become important to BWA deployments

synchronization and channel equalization methods for OFDM. Synchronization can also include frequency synchronization as well as timing synchronization.

At higher frequencies, LOS is a must. This requirement eases the effect of multi-path, allowing for wide channels, typically greater than 10 MHz in bandwidth. This gives IEEE 802.16 the ability to provide very high capacity links on both the uplink and downlink. For sub-11–GHz, NLOS capability is a requirement. The original IEEE 802.16 MAC was enhanced to accommodate different PHY layers and services, which address the needs of different environments. The standard is designed to accommodate either TDD or FDD deployments, allowing for both full and half-duplex terminals in the FDD case.

IEEE 802.16 MAC Layer

The main focus of the MAC layer is to manage the resources of the air link in an efficient manner. The MAC was designed for the point-to-multi-point wireless access environment. It supports higher layer or

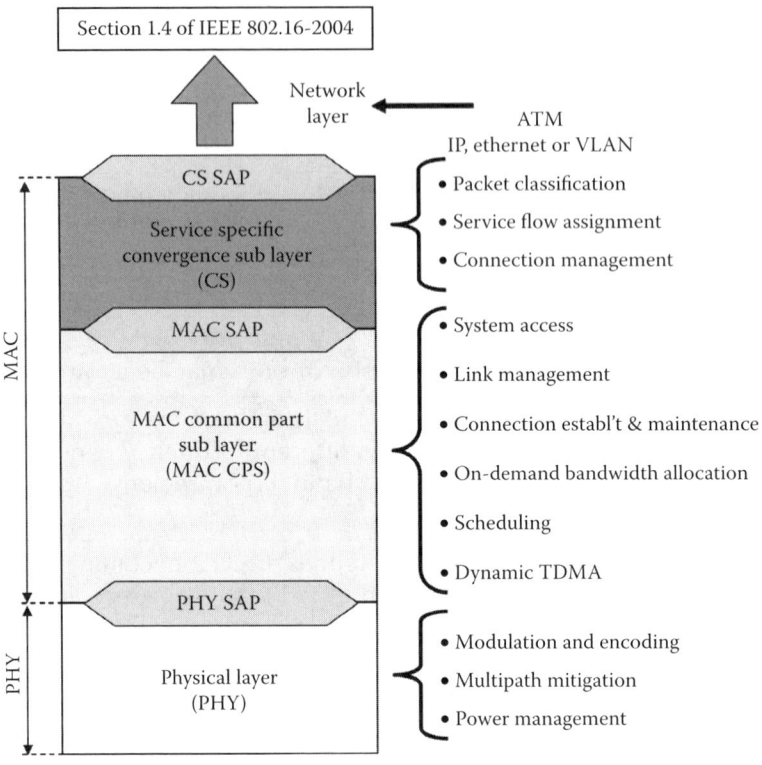

Figure 8.12 WiMAX PHY layer.

transport protocols such as ATM, Ethernet, or IP, and is designed to easily accommodate future protocols that have not yet been developed.

IEEE802.16 MAC is capable of supporting multiple physical layer specifications optimized for the frequency bands of the application. The standard includes a particular PHY layer specification broadly applicable to systems operating between 10 and 66 GHz.

IEEE802.16 MAC is designed for very high bit rates (up to 268 Mbps each way) of the truly broadband physical layer, while delivering ATM-compatible QoS.

IEEE802.16 MAC uses a variable-length protocol data unit (PDU) along with a number of other concepts that greatly increase the efficiency of the standard. Multiple MAC PDUs may be concatenated into a single burst to save PHY layer overhead.

Additionally, multiple service data units (SDUs) for the same service may be concatenated into a single MAC PDU, saving on MAC header

Table 8.2 IEEE 802.16a MAC Layer Features: Benefits

Features	Benefits
TDM/TDMA scheduled uplink/downlink frames	Efficient bandwidth usage
Scalable from one to hundreds of subscribers	Allows cost-effective deployments by supporting enough subscribers to deliver a robust business case
Connection oriented	Per-connection QoS; faster packet routing and forwarding
QoS support	Low latency for delay-sensitive services (TDM voice, VoIP); Optimal transport for VBR traffic (e.g., video);· data prioritization
Automatic repeat request (ARQ)	Improves end-to-end performance by hiding RF-layer-induced errors from upper-layer protocols
Support for adaptive modulation	Enables highest data rates allowed by channel conditions, improving system capacity
Security and encryption (Triple DES)	Protects user privacy
Automatic power control	Enables cellular deployments by minimizing self-interference

overhead. Fragmentation allows very large SDUs to be sent across frame boundaries to guarantee the QoS of competing services. And, payload header suppression can be used to reduce the overhead caused by the redundant portions of SDU headers.

The MAC layer uses a self-correcting bandwidth request/grant scheme that eliminates the overhead and delay of acknowledgments, while allowing better QoS handling than with traditional acknowledged schemes.

Terminals have a variety of options for requesting bandwidth depending on the QoS and traffic parameters of their services. They can be polled individually or in groups. They can steal bandwidth already allocated to make requests for more. They can signal the need to be polled, and they can piggyback requests for bandwidth.

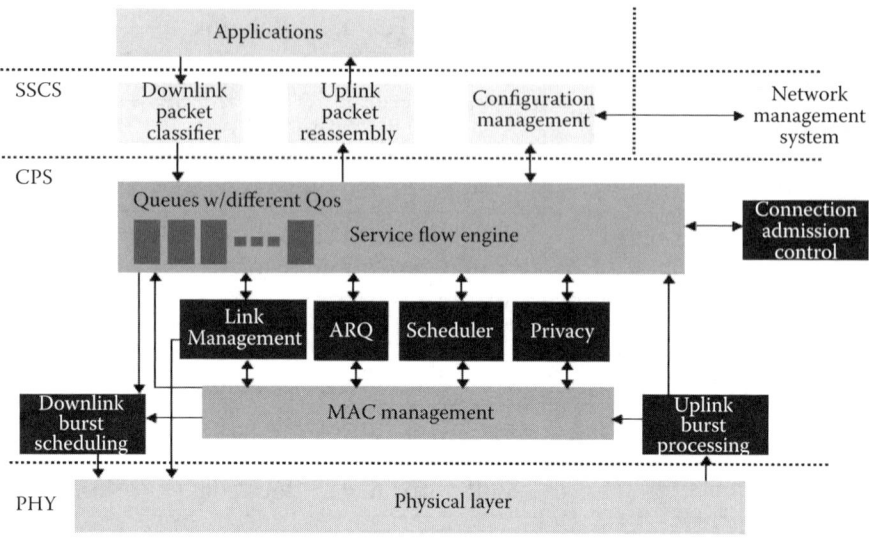

Figure 8.13   WiMAX MAC layer.

Overview

The following list summarizes the main features of the MAC layer:

It provides point-to-multi-point network access.
It supports metropolitan area networks.
It is connection oriented.
It supports difficult user environments:
■ High bandwidth, hundreds of users per channel
■ Continuous and bursty traffic
■ Very efficient use of spectrum
It has a protocol-independent core (ATM, IP, Ethernet, etc.).
It provides a balance between stability of contentionless and efficiency of contention-based operation.
It offers flexible QoS.
It supports multiple 802.16 PHY layers, i.e., support both TDD and FDD in the PHY layer.

Sublayers

The MAC layer consists of three sublayers.

The service-specific convergence sublayer (SSCS) provides an interface to the upper layer entities through a CS service access point (SAP).

The MAC common part sublayer (CPS) provides the core MAC functions, including uplink scheduling, bandwidth request and grant, connection control, and automatic repeat request (ARQ).

The privacy sublayer (PS) provides authentication and data encryption functions.

Smart Antenna Support

Smart antennas are being used to increase the spectral density (that is, the number of bits that can be communicated over a given channel in a given time) and to increase the signal-to-noise ratio (SNR) for both WiMAX solutions and for other wireless technologies such as Wi-Fi. The 802.16-2004 standards, owing to performance and technology, support several adaptive smart antenna types as discussed in the following text:

- Receive spatial diversity antennas: These entail more than one antenna receiving the signal. The antennas need to be placed at least half a wavelength apart to operate effectively; the wavelength can be derived by taking the inverse of the frequency. For example, for a 2.5 GHz carrier the wavelength would be 0.13 m; hence, the distance between antennas would be 0.065 m, or about 2.5 in. Maintaining this minimum distance ensures that the antennas are incoherent, that is, they will be impacted differently by the additive/subtractive effects of signals arriving by multiple paths.

- Simple diversity antennas: These detect the signal strength of the multiple (two or more) antennas attached and switch that antenna into the receiver. The likelihood of getting a strong signal depends on number of incoherent antennas available to choose from.

- Beam-steering antennas: These shape the antenna array pattern to produce high gains in the useful signal direction or notches that reject interference. High antenna gain increases the signal, noise, and rate. The directional pattern attenuates the interference out of the main beam. Selective fading can be mitigated if multi-path components arrive with sufficient angular separation.

■ Beam-forming antennas: These allow the area around a base station to be divided into sectors, allowing additional frequency reuse among sectors. The number of sectors can range from as few as 4 to as many as 24. Base stations that intelligently manage sectors have been used for a long time in mobile-service base stations.

Flexible Channel Bandwidth

As the distance between a subscriber and the base station increases, or as the subscriber starts to move by walking or driving in a car, it becomes more of a challenge for him or her to transmit successfully back to the base station at a given power level. For power-sensitive platforms such as laptop computers or handheld devices, it is often not possible for them to transmit to the base station over long distances if the channel bandwidth is wide.

Unlike 802.11 (fixed 20 MHz channel bandwidth) and 3G (limited channel bandwidth of 1.5 MHz), IEEE 802.16-2004 and IEEE 802.16e standards have flexible channel bandwidths between 1.5 and 20 MHz to facilitate transmission over longer ranges and to different types of subscriber platforms. In addition, this flexibility of channel bandwidth is also crucial for cell planning, especially in the licensed spectrum.

Initial applications for 802.16x products are likely to focus on wireless delivery of broadband Internet connectivity by communications providers, especially in rural areas. There is also significant potential for delivery of broadband to rural areas by private organizations.

RF Signals

Multiplexing Technology

WiMAX uses OFDM, a multicarrier technique that allows broadband transmission in a mobile environment with fewer multi-path effects than a single signal with broad bandwidth modulation.

Orthogonal Frequency Division Multiplexing

OFDM is a multicarrier transmission technique that has been recently recognized as an excellent method for high-speed bidirectional wireless data communication. Its history dates back to the 1960s, but it has recently become popular because economical integrated circuits that can perform the necessary high-speed digital operations have become available.

OFDM effectively squeezes multiple modulated carriers tightly together, reducing the required bandwidth but keeping the modulated signals orthogonal so they do not interfere with each other. Today, the technology is used in such systems as asymmetric digital subscriber line (ADSL) as well as wireless systems such as IEEE 802.11a/g (Wi-Fi) and IEEE 802.16 (WiMAX). It is also used for wireless digital audio and video broadcasting.

It is based on frequency division multiplexing (FDM), which is a technology that uses multiple frequencies to simultaneously transmit multiple signals in parallel. Each signal has its own frequency range (subcarrier), which is then modulated by data. Subcarriers are separated by guard bands to ensure that they do not overlap. These subcarriers are then demodulated at the receiver by using filters to separate the bands.

OFDM is similar to FDM but achieves more spectral efficiency by spacing the subchannels much closer together (until they are actually overlapping). This is done by finding frequencies that are orthogonal, which means that they are perpendicular in a mathematical sense, allowing the spectrum of each subchannel to overlap another without interfering with it. The effect of this is that the required bandwidth is greatly reduced by removing guard bands and allowing signals to overlap. To demodulate the signal, a discrete Fourier transform (DFT) is needed. Fast Fourier transform (FFT) chips make this a relatively easy operation.

OFDMA allows some subcarriers to be assigned to different users. These groups of subcarriers are known as *subchannels*. Scalable OFDMA allows smaller FFT sizes to improve performance (efficiency) for lower-bandwidth channels. This applies to IEEE 802.16-2004, which can now reduce the FFT size from 2048 to 128 to handle channel bandwidths ranging from 1.25 to 20 MHz. This allows subcarrier spacing to remain constant independently of bandwidth, which reduces complexity while allowing larger FFT for increased performance with wide channels.

Another advantage of OFDM is its resilience to multi-path, which is the effect of multiple reflected signals hitting the receiver. This results in interference and frequency-selective fading, which OFDM is able to overcome by utilizing its parallel, slower-bandwidth nature. This makes OFDM ideal to handle the harsh conditions of the mobile wireless environment.

OFDM's high spectral efficiency and resistance to multi-path make it an extremely suitable technology to meet the demands of wireless data traffic. This has made it not only ideal for such new technologies such as WiMAX and Wi-Fi, but it is also currently one of the prime technologies being considered for use in future fourth-generation (4G) networks.

Modulating Technology

A WiMAX provider can meet a wide variety of needs with a single distribution point by providing flexible service and rate structures to its customers. Depending on specific demand, it is possible for providers to offer a wide variety of standard and custom service offerings. This is possible because of the inherent modulation techniques used in IEEE 802.16. These techniques are also the basis of communications for systems such as cable modems, DSL modems, CDMA, 3G, Wi-Fi (IEEE 802.11), and futuristic 4G.

Modulation is the process by which a carrier wave is able to carry a message or digital signal (series of ones and zeroes). There are three basic methods for this: amplitude shift keying (ASK), frequency shift keying (FSK), and phase shift keying (PSK). Higher orders of modulation allow us to encode more bits per symbol or period (time).

In the case of WiMAX, ASK and PSK can be combined to create quadrature amplitude modulation (QAM), in which both the phase and amplitude are changed. The receiver then receives this modulated signal, detects the shifts, and demodulates the signal back into the original data stream.

The modulation scheme is dynamically assigned by the base station, depending on the distance to the client, as well as weather, signal interference, and other transient factors. This flexibility further enables service providers to tailor the reach of the technology to the needs of individual distribution areas, allowing WiMAX service to be profitable in a wide variety of geographic and demographic areas.

802.16 supports adaptive modulation, which allows it to automatically increase effective range, when necessary, at the cost of decreasing throughput. Higher-order modulation (e.g., 64QAM provides high throughput at submaximum range, whereas lower-order modulation (e.g., 16QAM) provides lower throughput at higher range, from the same base station.

Table 8.3   Modulation and Encoding Schemes

Channel Size (MHz)	Bit Rate (Mbps) QPSK	Bit Rate (Mbps) 16QAM	Bit Rate (Mbps) 64QAM
20	32	64	96
25	40	80	120
28	44.8	89.6	134.4

Adaptive Modulation

The use of adaptive modulation allows a wireless system to choose the highest-order modulation, depending on the channel conditions. In the case of WiMAX, either PSK or QAM is typically employed to increase the data throughput.

Different-order modulations allow sending more bits per symbol, thus achieving higher throughputs or better spectral efficiencies. However, it must also be noted that when using a modulation technique such as 64QAM, better SNRs are needed to overcome any interference and maintain a certain bit error ratio (BER).

As the range is increased, modulation is lowered (in other words, BPSK), but when range reduces, higher-order modulations such as QAM can be utilized for increasing throughput. In addition, adaptive modulation allows the system to overcome fading and other interferences. The modulated signals are then demodulated at the receiver, where the original digital message can be recovered.

Both QAM and QPSK are modulation techniques used in IEEE 802.16 (WiMAX). They are also used in IEEE 802.11 (Wi-Fi), and 3G (WCDMA/HSDPA) wireless technologies. The use of adaptive modulation allows wireless technologies to optimize throughput, yielding higher throughputs while also covering long distances.

Duplexing Technology

Duplexing refers to the process of creating bidirectional channels for uplink and downlink data transmission.

TDD and FDD are both supported by the 802.16-2004 standards. FDD, unlike TDD, requires two channel pairs that are separated to

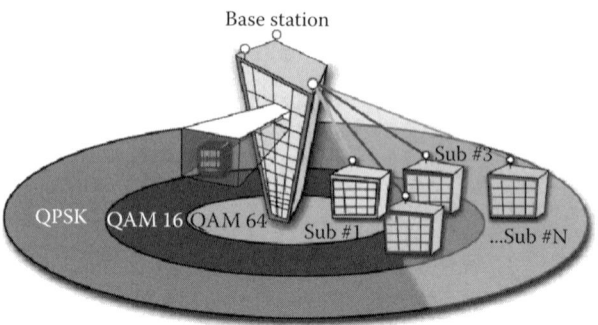

Figure 8.14 Adaptive modulation.

minimize interference, one for transmission and the other for reception. Most FDD bands are allocated to voice, because the bidirectional architecture of FDD allows voice to be handled with minimal delays whereas TDD is more efficient for IP or data.

FDD, however, adds additional components to the system and therefore increases costs. FDD is also used in 3G networks, which operate at a known frequency and are designed for voice applications. FDD has limitations for data throughput. As network traffic increases or decreases, the geographic area covered by the transmitter may shrink or grow, a phenomenon called *cell breathing.*

IEEE 802.16 specifies both FDD and TDD options which are considered in the following text.

FDD and TDD

The following are the chief characteristics of FDD:

- Support for legacy services
- Symmetrical traffic only
- Inflexible deployment
- Lower efficiency (especially HD-FDD)
- Necessity of a guard band

The following are the chief characteristics of TDD:

- Efficient for IP-based systems
- Both symmetrical and asymmetrical
- Flexibility; single band required

Table 8.4   TDD versus FDD

	TDD	*FDD*
Description	A duplexing technique used in license-exempt solutions and which uses a single channel for both the uplink and downlink	A duplexing technique utilized in licensed solutions that uses a pair of spectrum channels, one for the uplink and another for the downlink
Advantages	Enhanced flexibility because a paired spectrum is not required; Easier to pair with smart antenna technologies; Asymmetrical	Proven technology for voice; designed for symmetrical traffic; does not require guard time
Disadvantages	Cannot transmit and receive at the same time	Cannot be deployed where spectrum is unpaired; spectrum usually licensed; higher cost associated with spectrum purchase
Usage	Bursty, asymmetrical data applications; Environments with varying traffic patterns; In which RF efficiency is more important than cost	Environments with predictable traffic patterns; Where equipment costs are more important than RF efficiency

- Highly efficient, twice the bandwidth
- Adaptability with advance signal processing (i.e., AAS)

Both FDD and TDD systems can coexist, for example, in adjacent bands. Recommendations are provided in IEEE, European Telecommunications Standards Institute (ETSI), and CEPT documents. Coexistence has been demonstrated on the field in many deployment cases.

WiMAX Architecture

A wireless MAN based on the WiMAX air interface standard is configured in much the same way as a traditional cellular network with strategically located base stations using a point-to-multi-point architecture to deliver services over a radius of up to several miles, depending on frequency, transmit power, and receiver sensitivity. In areas with high population densities, the range will generally be capacity limited rather than range limited, owing to limited bandwidth.

The base stations are typically backhauled to the core network by means of fiber or point-to-point microwave links to available fiber nodes or via leased lines from an existing wireline operator. The range and NLOS capability make the technology equally attractive and cost-effective in a wide variety of environments. The technology was envisioned from the beginning as a means of providing wireless last mile broadband access in the MAN with performance and services comparable to or better than traditional DSL, cable, or T1/E1 leased line services.

The technology is expected to be adopted by different incumbent operator types, for example, wireless internet service providers (WISPs), cellular operators (CDMA and WCDMA), and wireline broadband providers. Each of these operators will approach the market with different business models based on their current markets and perceived opportunities for broadband wireless as well as different requirements for integration with existing (legacy) networks. As a result, 802.16 network deployments face the challenging task of needing to adapt to different network architectures while supporting standardized components and interfaces for multi-vendor interoperability.

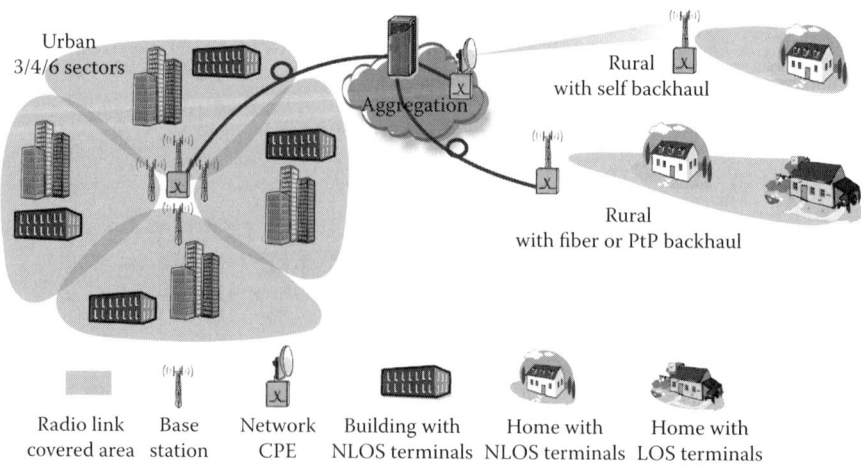

Figure 8.15   WiMAX in rural and urban landscapes.

Deployment Best Practices

Three steps are required to complete the project of WiMAX deployment and minimize the risk of costly modifications. The three steps are as follows:

Defining the Requirements

Matching data density requirements to base station capacity is the key to an optimized deployment. For capacity-limited deployment scenarios, it is necessary to deploy base stations spaced apart sufficiently to match the expected density of end users. Data density is an excellent metric for matching base station capacity to market requirements. Demographic information, including population, households, and businesses per square kilometer or per square mile, is readily available from a variety of sources for most areas. With this information and the expected services to be offered, along with the expected market penetration, data density requirements can be calculated easily. This six-step process is summarized as follows:

1. Target market segment
2. Area demographics
3. Services to be offered
4. Expected market take rate
5. Expected number of customers
6. Required data density Mbps per square kilometer

Once the target market segment is decided, a table can be created similar to the example shown (Table 8.5). Values from this table are used to generate the graphs shown in Figure 8.16.

Data Density Requirements Based on Demographics: Expected Residential and SME Market Penetration

Based on the service definitions in Table 8.5, the typical range of data density requirements based on area demography, i.e., an urban, suburban, or rural environment, for the metropolitan area under consideration is derived. Again, a table can be created (Table 8.6). After considering the oversubscription factor, which depends on the business model, the required data density is decided. A sample deployment for all three demographics, i.e., urban, suburban, or rural, is shown (Table 8.7).

Table 8.5   Sample Table Representing Market Segment

Customer Type	Service Description	Overbooking Factor
Residential	384 kbps average	20:1
Residential VOIP (20 percent of users)	128 kbps average	4:1
SME Premium (25 percent)	1.0 Mbps CIR, 5 Mbps PIR	1:1 (CIR)
SME Regular (75 percent)	0.5 Mbps CIR, 1 Mbps PIR	1:1 (CIR)

Note: CIR = committed Internet data rate, PIR = peak Internet data rates.

Table 8.6   Sample Table Representing Typical Data Rate Requirements

	Urban	Suburban	Rural
Residential density penetration	4000 to 8000 5 to 10 percent	800 to 1500 5 to 10 percent	200 to 600 5 to 10 percent
SME density penetration	400 to 600 2 to 5 percent	50 to 100 2 to 5 percent	10 to 30 2 to 5 percent
Data density range	10 to 40 Mbps/sq km	2 to 7 Mbps/sq km	0.5 to 2 Mbps/sq km

Table 8.7   Sample Table with Demographics for Deployment

	Urban	Suburban	Rural
Geographic area to be covered	60 sq km	120 sq km	200 sq km
Expected number of residential customers	30,000	20,000	5,000
Expected number of SME customers	1,500	500	150
Required data density	29 Mbps/sq km	5.9 Mbps/sq km	1.0 Mbps/sq km

Table 8.8   Sample Table with Possible Solutions

Band	Duplex	Channels	Spectrum Required	Terrain Condition	
2.5 GHz	TDD	3	15 MHz	Rural	NLOS
3.5 GHz	FDD	3	21 MHz	Rural	NLOS

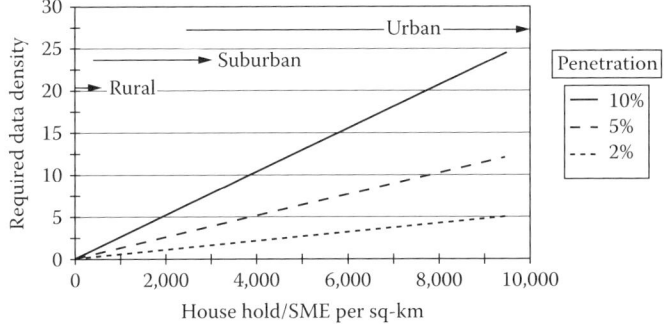

Figure 8.16   Data density versus base station distance.

Once we have the required data density for the area, we go to the second step, which is to understand the terrain and plan the best possible layout.

Site Survey

A site survey is done to identify points where base stations can be installed so as to provide the required data density in the most efficient way.

The site survey provides response to questions such as the type of base station or CPE that must be used and the modulation options that can be employed (depending on coverage area). Finally, based on these factors the profile of the system is finalized. An example with two choices is shown in Table 8.8.

The decision to select one of these profiles will depend heavily on factors such as cost, maintenance, government regulations, etc.

Once the profile to be installed is decided on, the distance between base stations to be deployed can be found using reference sheets that have radio characteristics such as the graph of data density to base station distance shown in Figure 8.16.

Physical Deployment

Information about site and base station distance allows planners to select locations for the base stations. A complete site plan with points showing proposed base stations is provided to the field team for site preparation and base station commissioning.

Deployment Stages

Initially, IEEE 802.16 standards-based networks will likely target fixed-access connectivity to unserved and underserved markets in which wireline broadband services are insufficient to fulfill the market need for high-bandwidth Internet connectivity.

Prestandards implementations exist today that are beginning to address this fixed-access service environment. Standardization will help accelerate these fixed-access solutions by providing interoperability among equipment and economies of scale resulting from high-volume standards-based components.

As IEEE 802.16 solutions evolve to address portable and mobile applications, the required features and performance of the system will increase. Beyond fixed-access service, even larger market opportunities exist for providing cost-effective broadband data services to users on the move. Initially, this includes portable connectivity for customers who are not within reach of their existing fixed broadband or WLAN service options. This type of service is characterized by access that is unwired but stationary in most cases, albeit with some limited provisions for user mobility during the connection.

In this manner, 802.16 can be seen as augmenting coverage of 802.11 for private and public service networks and cost-effectively extending hot spot availability to wider ranges of coverage. Based on this described capability, this phase of deployment is referred to as *nomadic*, or *portability with simple mobility*.

The next phase of functionality, known as *full mobility*, provides incremental support for low-latency, low-packet-loss real-time handovers between base stations at speeds of 120 km/hr or higher, both within a network and between networks. This will deliver a rich end-user experience for high-quality multimedia applications.

Network Topology

Today, there are technologies emerging that take advantage of LOS transmission capabilities long thought out of date. Whereas LOS

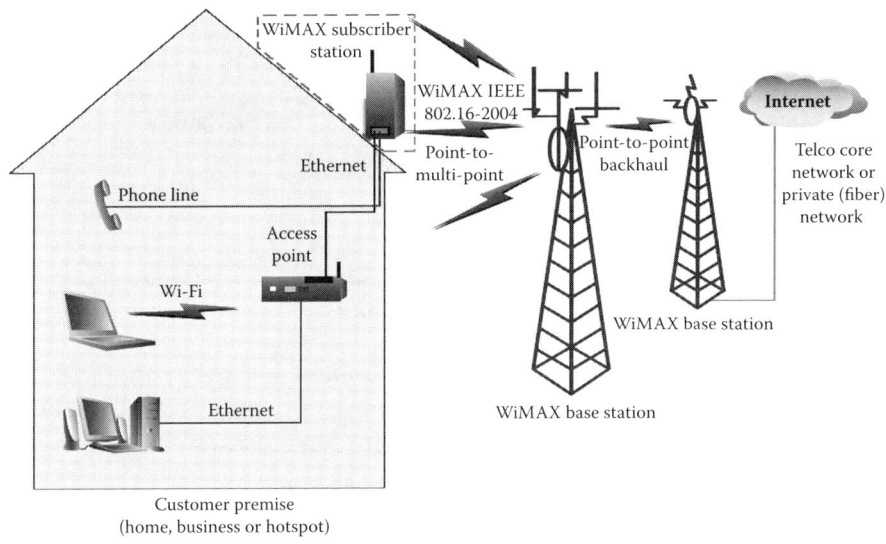

Figure 8.17   WiMAX network topology.

technology was more often than not point to point, today's advances allow for point to multi-point, providing a much more cost-effective service. Some of these technologies can even support obstructed transmission paths, common in typical communities.

Terrestrial Fixed Wireless Access (FWA)

The current status is adequate for widespread commercial applications in the frequency bands up to about 40 GHz and for the emerging applications up to about 60 GHz. A variety of device structures exist, and new variations will proliferate in the contest for optimal solutions in the various circuit applications.

P-P Networks

P-P fixed wireless networks are commonly deployed to offer high-speed dedicated links between high-density nodes in a network. Such systems are cost-effective and can be deployed easily. Moreover, as a large part of a wireless network's cost is not incurred until the CPE is installed, the network service operator can time capital expenditures to coincide with the signing up of new customers.

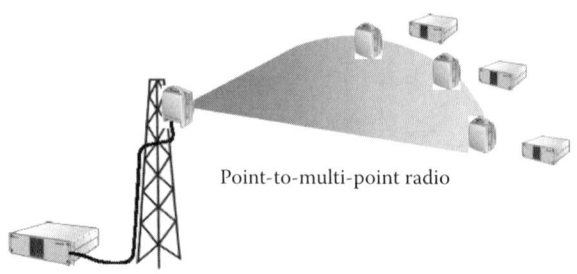

Figure 8.18 WiMAX point-to-point topology.

P-P systems provide an effective last mile solution for the existing service provider and can be used by competitive providers to deliver services directly to end users. Benefits can be summarized as follows:

- Lower entry and deployment costs
- Ease and speed of deployment (rapid development with minimal disruption to the community and the environment)
- Fast realization of revenue (as a result of rapid deployment)
- Demand-based build-out (scalable architecture employing open industry standards ensuring services and coverage areas can be easily expanded as customer demand warrants)
- Cost shift from fixed to variable components
- No stranded capital when customers churn
- Cost-effective network maintenance, management, and operating costs

P-MP Networks

P-MP is a concept in which multiple subscribers can access the same radio platform, utilizing both a multiplexing method and queuing. More recent advances in P-MP technology offer service providers a method of providing high-capacity local access that is less capital intensive and faster to deploy than wireline, and that is able to offer a combination of applications.

P-MP implementations are emerging in several bands above 20 GHz, up to about 40 GHz. These consist of a complex TDM hub or base station using sectoral antennas and TDMA subscriber stations using parabolic antennas.

Most P-MP systems use a simple modulation method, e.g., QPSK, but higher-level modulation methods are also used in some systems,

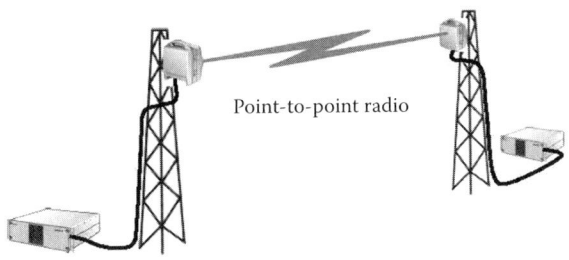

Point-to-point radio

Figure 8.19   WiMAX point-to-multi-point topology.

e.g., 64QAM. There is a trade-off between modulation method, interference tolerance and link length: more advanced P-MP system capabilities, including even higher level modulation methods.

Local Multi-Point Distribution System

LMDS is a unique wireless access system whose purpose is to provide broadband access to multiple subscribers in the same geographic area. LMDS, although operating in the microwave frequency band and utilizing similar radio technology as a P-P microwave system, enables an operator to handle more subscribers, or rather Mbps/sq km, than a microwave P-P system using the same amount of RF spectrum.

LMDS utilizes microwave radio as the fundamental transport medium. It is not a fundamentally new technology but an adaptation of existing technology for a new service implementation. The new service implementation allows multiple users to access the same radio spectrum.

Network Architecture

As different LMDS system operators offer different services and have different legacy systems and business strategies, the system architectures they use also differ. LMDS, being a LOS transmission system, requires a central antenna at a relatively high point, such as on top of a tall building or tower, to serve the surrounding geographic area. In addition, LMDS operates in the 28 to 31 GHz frequency band, which results in a relatively short wavelength; therefore, the transmission distance is limited.

In practice, the transmission distance depends on the modulation method employed as well as the type of geographic area, particularly

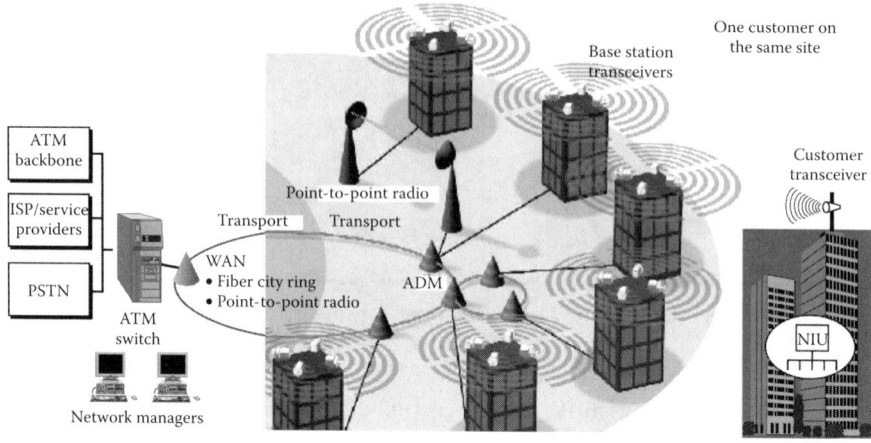

Figure 8.20 WiMAX Local Multi-Point Distribution System (LMDS).

with respect to expected heavy rain. The more efficient the modulation method, the shorter the transmission distance. Similarly, in an area where periodic heavy rain can be expected, coverage must be designed for a smaller geographic area. Moreover, LMDS is probably more suitable for use in suburban areas than in high-density urban areas, where numerous tall buildings and towers can easily obstruct LOS communications.

The most common architectural elements used in LMDS systems are described in the following subsections.

LMDS Cell

The base station associated with mobile wireless applications is commonly referred as an *LMDS node*, and the surrounding area that this node can possibly serve is called an *LMDS cell*. An LMDS node can use virtually any type of connection. It is possible for an LMDS node to be connected to an earth station that receives satellite television feed, enabling the technology to function as a wireless cable operator that can also provide voice and data services.

At minimum, each node will be connected either directly or indirectly to a network operation center (NOC). The series of base stations or nodes within a geographic area operated by a company and controlled by one or more NOCs will function as an entity for billing, traffic, management, registration, and authentication operations.

Access Method

The actual method of LMDS access to subscribers can be expected to vary from vendor to vendor. Taking advantage of previously developed cellular technologies, it is possible for TDMA, FDMA, and CDMA to be used by an LMDS operator.

Although it is possible to use FDMA, TDMA, or CDMA as an access method, the current implementations of LMDS use the first two methods. In addition, it is quite possible that an LMDS operator could employ different access technologies at different nodes, similar to the manner in which Advanced Mobile Phone Service (AMPS) and Digital Advanced Mobile Phone Service (D-AMPS) coexist within a geographic area.

Base Station

A typical LMDS base station or node employs an on-site methodology, under which both the connection to the network and outdoor microwave antenna are located within the same structure. Because microwave transmission is an LOS communications method operating within the constraints of the antenna, signals are focused within a certain range. A typical transmit and receive sector antenna will provide service over a 15°, 30°, 45°, or 90° beam width. Thus, the base station can be expected to have multiple antennas mounted on top of the structure.

Network Interface Unit

The network interface unit (NIU) resides at the customer premise and is the opposite side of the base station wireless link. NIU provides a gateway between the RF communications that provide a transport service to the base station and in-building communications facilities, such as a LAN, PBX, communications front-end processor, and other devices.

Mesh Networks

Mesh topologies provide a flexible, effective, reliable, economical, and portable architecture that can move data between nodes efficiently while maintaining balanced traffic along the network.

Mesh networks are wireless data networks composed of two or more autonomous, self-organizing nodes. The nodes are similar to traditional wireless transmitter receivers (similar to a base station or wireless network card) but have additional intelligence built in that enables them to act as minirouters for the network. The nodes are installed throughout a large area (such as a colony or a school campus). Each node then transmits a low-power signal capable of reaching neighboring nodes, each of which in turn transmits the signal to the next node, the process being repeated until the data arrives at its destination. By adding the capacity of each node to route packets to other nodes in the network, meshes can extend the range of wireless technologies such as 802.11b/g and provide low-cost coverage of a geographic area using a single broadband connection.

Connections between nodes are made on demand, and packets are routed across the network either using predetermined routing tables (proactive routing) or routes generated on demand by the network (reactive routing). Properly configured mesh networks should be self-healing: if a node goes down, the remaining nodes in the network can reconfigure their routes to work around the failure.

Mesh networks also scale up and out in very small size increments. A project can start with a few nodes and then scale up a single node at a time. The mesh network is based on multi-hop topology, which has many advantages as well as a few disadvantages.

Multi-Hop Topology

Up to this point, wireless systems have been represented as consisting of base stations or APs that feed a collection of end systems. But more complex wireless systems are also feasible. For example, a simple case would consist of an 802.16 wireless-access system with a rooftop antenna wired to an 802.11 AP inside the building with which end systems communicate.

In a more complex example, instead of feeding an 802.16 base station directly with fiber, a link from another 802.16 base station can be used as a feed. A multi-hop 802.16 system requires careful channel selection so that the feeder signal from the first base station does not interfere with the distribution signal sent from the end base station; nevertheless, such a system might obviate considerable construction of fiber links when serving sparsely populated areas.

Multi-hop is a better topology than single-hop and directional last mile alternatives. It is more robust than single-hop networks because they are not dependent on the performance of a single node for operation. In a single-hop network, if the node goes down, so does the network. In mesh-network architecture, if the nearest node goes down or if localized interference occurs, the network continues to operate; data is simply routed along an alternate path.

Also, multi-hop networks use available bandwidth efficiently. In a single-hop network, devices must share a node, because of which several devices attempt to access the network at once, causing a traffic jam and system slow down. By contrast, in a multi-hop network, many devices can connect to the network at the same time through different nodes, without necessarily degrading system performance. The shorter transmission ranges in a multi-hop network limit interference, allowing simultaneous, spatially separated data flows. To deploy a multi-hop network cost-effectively, however, service providers need a large initial subscriber base.

Technical considerations such as network latency experienced by the end systems, which keeps escalating with increase in the number of hops, may limit the effectiveness of this approach by adversely impacting some applications such as voice. In designing multi-hop links that will carry latency-sensitive traffic, an acceptable latency budget needs to be defined, after which the latency added for each relay point has to be computed. Generally, a few hops can be tolerated without unacceptable degradation to voice traffic (i.e., latencies worse than what is experienced by cell users or general VoIP users today). The specific limits will depend on the specifications of the actual equipment, however.

Another limiter to multi-hop deployments is that the traffic from outlying base stations will fan in eventually to a common link to the high-performance fiber backbone, and the total capacity of the common links can eventually limit total system performance.

Mesh Design

An even more complex system architecture that may be appropriate for some deployments is the one with a mesh design. Such designs are still in the research stage today but will likely be incorporated into future standards. In a mesh design, all or at least very many endpoint nodes also act as relay points to other endpoint nodes. Essentially, traffic to many users is carried through radios at their neighbors' homes. This reduces the number of explicit base stations that are needed by turning every endpoint into a kind of mini base station. For cases

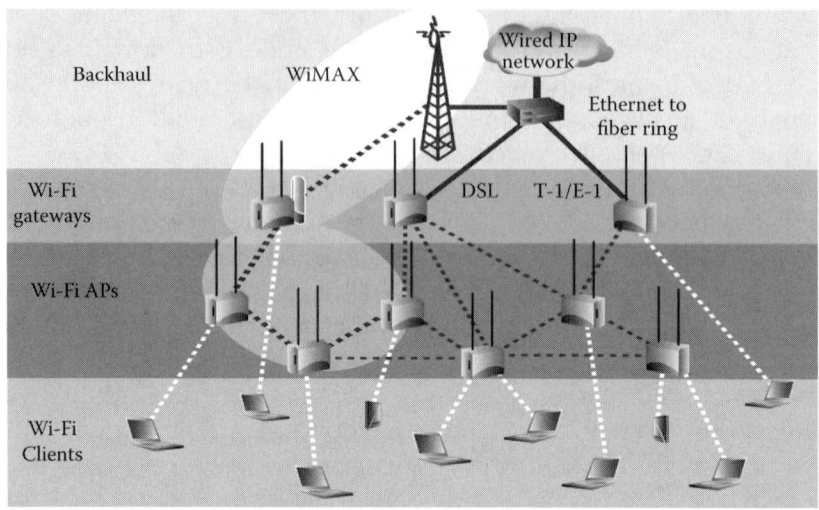

Figure 8.21   Mesh network.

where the endpoints to be served are very sparse, this may reduce deployment costs by eliminating central base stations in favor of adding small incremental costs to many user stations. For more dense deployments, mesh systems can increase reliability and capacity because there may be many traffic paths back to the Internet.

Mesh networks are a highly innovative extension of wireless networking technology. Most deployments and commercial efforts to date have focused on military and emergency services applications, but the potential exists for mesh networks to be used to extend the reach of broadband Internet connections.

Mesh Types

There are two types of mesh networks that are based on network topology, referred as *full mesh* and *partial mesh*.

In full mesh networks, every node has a circuit connecting it to every other node in the entire network. This type of network can be expensive to build, but it offers the greatest degree of network redundancy.

In partial mesh networks, only some nodes operate in a full mesh arrangement, and other nodes connect to perhaps just one or two others in the network. This type of network is less expensive to implement, but offers incomplete redundancy for the network.

Mesh Advantage

An advantage of mesh topology is the ability of the deployment to navigate around a large obstacle, such as a mountain or tall buildings. Such obstacles can block a subscriber from reaching a base station. In a mesh network, blocked subscribers can get to the base station indirectly by going through other nodes. Even a small amount of meshing can greatly improve a base station's coverage if sufficient small nodes are in place.

Mesh networks, unlike direct LOS implementations, can adapt to changes in the network, making them more effective. In this topology, nodes can be readily added or removed, and their location can also be changed. As people become more mobile and wireless capabilities are included in new classes of devices, future business and home networks need to adapt or self-configure to these changes.

Mesh networks lower costs for the operator because users already have a client (such as a laptop with embedded Wi-Fi technology).

Mesh networks provide greater redundancy and can be used for traffic balancing. In dense networks, such as crowded offices or apartments, each device can have many neighbors, creating multiple paths between two communicating devices. In the presence of localized interference, a multi-hop network can route data along an alternate path. If only one node requires a large amount of bandwidth, then the network can dynamically route traffic to other network nodes, avoiding the congested node.

Mesh Drawbacks

Mesh networks do not conform to standards at this time. IEEE has begun work on a mesh standard (IEEE 802.11s), but full, approved standards are not expected until late 2006 or early 2007. Current back-end implementations of Wi-Fi/mesh infrastructures are based on proprietary solutions.

Mesh networks can be slow, because latency (the time it takes for a packet of information to cross a network connection from sender to receiver) increases with each network hop.

Mesh networks are also inherently noisy, because wireless mesh network links are multidirectional broadcasters and can pick up extraneous signals.

Scalability issues could arise because mesh networks involve a high degree of information routing between nodes.

Mesh Applications

There is no limit to innovative applications of mesh networks. For a basic understanding, three primary applications are described in the following subsections:

Fixed community networks that use meshes to share bandwidth
Self-assembling mobile networks for use in military and emergency
    services applications
Small-scale sensor networks

Community Networks

Many communities, in both the developed and developing world, may have limited broadband Internet access. Using mesh networks, these communities can share a single broadband connection among community members. In these community networks, a mesh of 802.11-based network nodes is set up and used to provide wireless coverage to a geographic area.

At least one of the nodes is connected to the Internet. The other nodes can be accessed via wireless cards in laptops or desktop computers and data is transmitted across the multi-hop mesh network to the Internet.

Military and Emergency Services

Mesh networks originated in the military sector as a means of providing ad hoc battlefield networks that are both self-assembling and self-healing, eliminating the vulnerability of a single network hub. In the emergency services sector, wireless mesh networks can be used to provide ad hoc networks between emergency services vehicles on location.

For example, a town might equip its police cruisers and fire department vehicles with laptop computers and mesh network nodes. At the scene of a fire, the nodes can self-assemble into an ad hoc network, providing communication between vehicles and to firefighters inside the building. If the vehicles are near enough to an equipped building, the ad hoc network may also be able to communicate back to the fire department or police headquarters to provide real-time information about the status of the emergency.

Sensor Networks

Most mesh network applications, especially in the commercial sector, focus on traditional PC-based computing. However, researchers are also interested in using mesh network technologies to create networks of autonomous sensors, which are small devices that can be installed in a variety of locations to provide readings on temperature, air quality, and other factors.

By incorporating a wireless chipset with mesh networking software, these sensors can become network aware. After they are installed and powered on, the sensors can join a mesh network and make their data accessible to others on the network. In many situations, both in buildings and outdoors, installing small mesh-enabled sensors in many locations will be far preferable to setting up network cabling to connect the sensors or (worse) manually collecting data from the sensors.

The effective combination of 802.16, 802.11, and fiber and wired infrastructure can support a wide variety of cost-effective Internet access, ranging from traditional office access to remote agricultural and community systems supporting rural development. It is worth noting that although the mobility benefits of wireless access may be viewed as being mainly of interest as a luxury for developed economies, they may be beneficial even in developing economies as a way of supporting such activities as in infrastructure development, outdoor industries such as farming, drilling, or mining, or even roving governmental services teams for medical outreach, etc.

Spectrum Issues

> Spectrum availability implies capacity for broadband wireless, which leads to the following interrelated cascading effects: a third pipe to the home; more broadband competition; better, cheaper, and more innovative services; and economic development and personal fulfillment.

The preceding lines clearly underscore how critical spectrum policies are for deciding the fate of a country's economy.

The radio spectrum is used by a wide variety of users ranging from consumer radio and television, to weather and aircraft radar, to data communications. Ranges of frequencies are assigned to various uses based on history, technical properties of the various frequencies, and other considerations. Because radio waves do not respect national

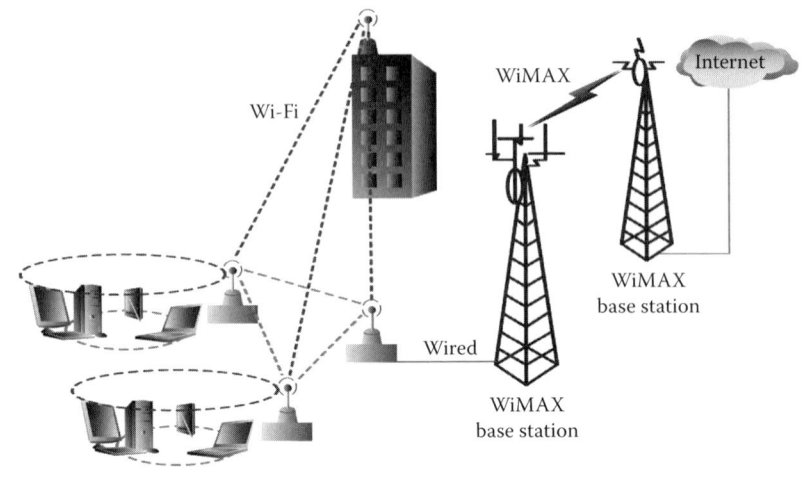

Figure 8.22   Mesh network as backhaul.

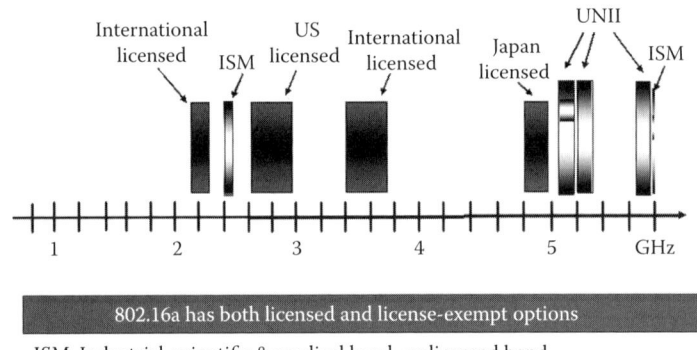

ISM: Industrial, scientific & medical band–unlicensed band

UNII: Unlicensed national information infrastructure band–unlicensed band

Figure 8.23   WiMAX spectrum.

borders, international agreements attempt to harmonize uses across multiple countries where conflicts could occur or where there is benefit in using common equipment.

Spectrum Management

To control how various bands are used, national regulations have commonly required licenses for operators to be permitted to use particular frequencies. Licensed bands, or radio channels, are most important to relatively high-power and longer-range uses of radio, in which a significant chance of interference between different radio users would exist were control not exercised. For example, television channels are allocated so that multiple broadcasters do not operate so near to one another that receivers would experience garbled reception.

Many national regulators have also defined as unlicensed or license-exempt bands for some types of radio use in cases, where requiring and enforcing licenses would be too cumbersome. To maintain order in such bands, there are generally rules limiting the power and other technical attributes that a radio operating in the bands have to confirm to. Because of the typical low power used, users are expected to be able to use such bands either without mutual interference or by managing any interference among themselves without governmental legal help.

There is general agreement among industry analysts that the traditional models of spectrum management are in need of reform. Most economists agree that the reform should seek to increase the ability of market forces to shape how spectrum is allocated and used. Traditional licenses that were encumbered with restrictions on the choice of technology, the services offered, their coverage, and the transferability of access rights have imposed a high opportunity cost for spectrum for many advanced communication services, while precluding the deployment of underutilized spectrum to higher-value uses. This has increased industry costs, reduced incentives to innovate, and slowed the deployment and adoption of new services.

Spectral Impact: WiMAX

The ability to operate a standardized solution in both a licensed and a license-exempt band is one of the benefits of WiMAX solutions for deployments around the world. Both licensed and license-exempt WiMAX solutions provide significant advantages over wired solutions in the areas of cost-effectiveness, scalability, and flexibility. The adoption of license-exempt and licensed WiMAX solutions is driven by the following additional benefits. The 802.16-2004 standard supports flexible RF channel bandwidths and reuse of these frequency channels as a way to increase network capacity. The standard also specifies support for TPC and channel-quality measurements as additional tools to support efficient spectrum use. The standard has been designed to scale up to hundreds or even thousands of users within one RF channel. Operators can reallocate spectrum through sectoring as the number of

Table 8.9   Global Licensed and License-Exempt Spectrum Bands

Country/Geographic Area	Bands — Licensed and License-Exempt
North America, Mexico	2.5 GHz and 5.8 GHz
Central and South America	2.5 GHz, 3.5 GHz, and 5.8 GHz
Western and Eastern Europe	3.5 GHz and 5.8 GHz
Middle East and Africa	3.5 GHz and 5.8 GHz
Asian Pacific	3.5 GHz and 5.8 GHz

subscribers grows. Support for multiple channels enables equipment makers to provide a means to address the range of spectrum use and allocation regulations operators have to cater to and comply within diverse international markets.

Licensed versus License Exempt

Governments around the world have established frequency bands available for use by licensed and license-exempt WiMAX technologies. Each geographic region defines and regulates its own set of licensed and license-exempt bands. To meet global regulatory requirements and allow providers to use all available spectrums within these bands, the 802.16-2004 standard supports channel sizes between 1.5 and 20 MHz.

What is sometimes overlooked is that each band provides a different set of advantages for different usage models. Each serves a different market need based on trade-offs between cost and QoS. License-exempt solutions and licensed solutions each offer certain advantages to providers. The availability of both allows providers and emerging markets to fulfill a variety of usage needs.

Licensed Spectrum

The 2.5 GHz band has been allocated in much of the world, including North America, Latin America, Western and Eastern Europe, and parts of Asia Pacific as a licensed band for WiMAX. Each country allocates the band differently, so the spectrum allocated across regions can range from 2.6 to 4.2 GHz.

In the United States, the FCC has created the Broadband Radio Service (BRS), previously called the Multi-Channel Multi-Point Distribution System (MMDS), for wireless broadband access. The restructuring that followed has allowed for the opening of the 2.495 to 2.690 GHz bands for licensed solutions such as 2.5GHz in WiMAX. In Europe, the ETSI has allotted the 3.5 GHz band, originally used for wireless local loop (WLL), for licensed WiMAX solutions.

To deploy a licensed solution, an operator or service provider must purchase spectrum. The purchasing of spectrum is a cumbersome process. In some countries, filing the appropriate permits to obtain licensing rights may take months, whereas in other countries, spectrum auctioning can drive up prices and cause spectrum acquisition delays.

Benefits of Licensed WiMAX Systems

A WiMAX system operating in the licensed band has an advantage over a system operating in an unlicensed band in that it has a more generous downlink power budget and can better support indoor antennas. Another significant advantage is that the lower frequencies associated with licensed bands (2.5 GHz and 3.5 GHz) enable better NLOS and RF penetration.

The higher costs and exclusive rights to spectrum enable a more predictable and stable solution for large metropolitan deployments and mobile usage. A higher barrier to entrance, coupled with exclusive ownership of a band, enables service quality improvements and reduces interference.

However, licensed bands are not without interference issues. As service providers deploy more networks, they must contend with mutual interference originating from within their own network. Proper design and implementation can alleviate these problems. In summary, licensed solutions offer improved QoS advantages over license-exempt solutions.

Applications of Licensed WiMAX Systems

As licensed WiMAX solution offers better control across large areas, enhanced scalability, QoS, and flexibility for users on the move, mobility-related issues such as transmitting RF signals to and from a moving target are more easily addressed using a licensed solution. Further, licensed solutions use FDD. WiMAX-licensed solutions are suitable for the following applications:

■ Large-coverage, P-MP applications
■ Ubiquitous broadband mobile services

- When licensing enables control over the usage of spectrum and interference
- When cost is not the primary issue for choosing the technology, because the technology has been optimized for this application (other technologies such as 3G data overlays will cost more and have worse performance)
- When services and base station equipment can only be leased from a carrier or service provider

License-Exempt

The most commonly discussed unlicensed band, available virtually worldwide today, is in the vicinity of 2.4 GHz. This band is often called the Industrial, Scientific, Medical (ISM) band because its initial allocation was to allow radio emissions by various kinds of equipment. This is the band that is being used today for WLANs according to the IEEE 802.11b/g standards and which has been branded by an industry group as Wi-Fi.

Another commonly discussed set of bands are in the space between 5 GHz and 6 GHz, where the IEEE 802.11a standard is defined to operate. The unlicensed allocations in this band have been the subject of recent international harmonization efforts through the ITU at the 2003 World Radiocommunication Conference (WRC03).

The majority of countries around the world have embraced the 5 GHz spectrum for license-exempt communications. The 5.15 GHz and 5.85 GHz bands have been designated as license exempt in much of the world. Approximately 300 MHz of spectrum is available in many markets globally, and an additional 255 MHz of license-exempt 5 GHz spectrum is available in highly populated markets such as the United States.

Some governments and service providers are concerned that interference resulting from the availability of too many license-exempt bands could affect critical public and government communication networks, such as radar systems. These countries and entities have become active in establishing limited control requirements for 5 GHz spectrums. For example, the United Kingdom is currently introducing restrictions on certain 5 GHz channels and is considering enforcement of the use of the DFS (Dynamic Frequency Select) function.

One key point that needs emphasis is that unlicensed does not mean unregulated, and the various operators providing wireless services still need to maintain a no-interference working plan and a "good neighbor" attitude, along with ensuring efficient spectrum utilization.

Table 8.10 Advantages of Licensed and License-Exempt Solutions

Licensed Solution Advantages	License-Exempt Solution Advantages
Better quality of service	Fast rollout
Better non-line-of-sight (NLOS) reception at lower frequencies	Lower costs
Higher barriers for entrance	More worldwide options

Benefits of License-Exempt WiMAX Systems

The costs associated with acquiring licensed bands are leading many WISPs and vertical markets to consider license-exempt solutions for specialized markets, such as rural areas and emerging markets.

License-exempt solutions provide several key advantages over licensed solutions, including lower initial costs, faster rollout, and a common band that can be used in much of the world (see Table 8.10). These benefits are fueling interest and have the potential for accelerating broadband adoption. Service providers in emerging markets, such as developing countries or mature countries with underdeveloped areas, can reduce time to market and initial costs by quickly deploying a license-exempt solution without time-consuming permits or auctions. Even mature areas can benefit from license-exempt solutions.

Some service providers can use a license-exempt solution to provide last mile access for home, business, or backhaul or as a supplemental network backup for their licensed or wired networks. A license-exempt solution is regulated in terms of the transmission output power, although a permit is usually not required. A device or service can use the band at any time as long as output power is controlled adequately.

Providers who are particularly concerned about QoS, for example, may find that a licensed solution provides them with more control over the service. A service provider wanting to serve an emerging or underdeveloped market with a business class service can use a license-exempt solution, with proper network design including site surveys and specialized antenna solutions, to offer certain service level agreements (SLAs) for their specialized markets.

Applications of License-Exempt WiMAX Systems

License-exempt WiMAX solutions are focused on rural areas, emerging markets, underdeveloped areas or underserved areas, such as an isolated college campus or farm. Further, license-exempt solutions use TDD. WiMAX license-exempt solutions are suitable for the following applications:

- P-P, long-distance solutions in sparsely populated environments
- P-MP solutions in rural communities (including some developing countries)
- Areas with small RF in-band noise or where interference in the unlicensed band can be controlled within the geography, such as large enterprise campuses, military barracks, and shipyards
- Where cost is the major factor governing a decision between competing wireless technologies
- When ownership of equipment is an option to the end user

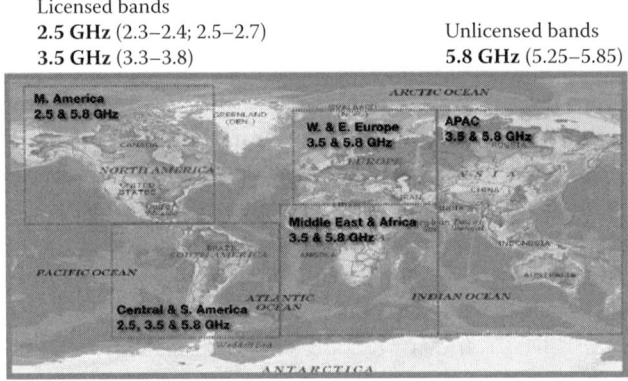

Figure 8.24   Global map — licensed and unlicensed bands.

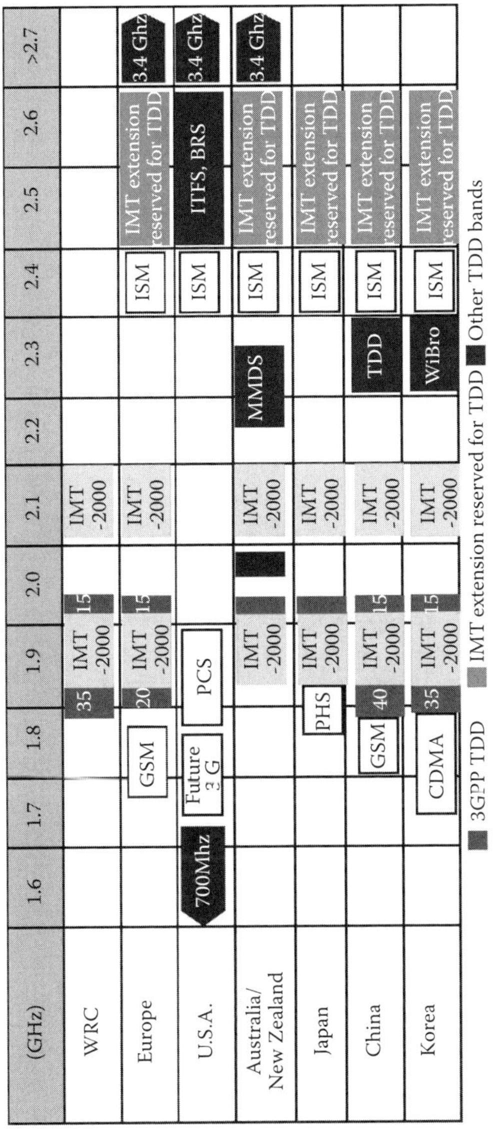

(GHz)	1.6	1.7	1.8	1.9	2.0	2.1	2.2	2.3	2.4	2.5	2.6	>2.7
WRC			35	IMT-2000 15		IMT-2000				IMT extension reserved for TDD		3.4 Ghz
Europe		GSM	20	IMT-2000 15		IMT-2000			ISM	IMT extension reserved for TDD		3.4 Ghz
U.S.A.	700Mhz	Future 3G		PCS					ISM	ITFS, BRS		3.4 Ghz
Australia/New Zealand				IMT-2000	15	IMT-2000	MMDS		ISM	IMT extension reserved for TDD		
Japan			PHS	IMT-2000		IMT-2000			ISM	IMT extension reserved for TDD		
China			GSM 40	IMT-2000 15		IMT-2000		TDD	ISM	IMT extension reserved for TDD		
Korea		CDMA	35	IMT-2000 15		IMT-2000		WiBro	ISM	IMT extension reserved for TDD		

■ 3GPP TDD   ▨ IMT extension reserved for TDD   ■ Other TDD bands

Figure 8.25   Global TDD spectrum allocation status.

Chapter 9

Surveying the Landscape

The wireless industry can never be accused of standing still. Doubts, if any, disappear when we look back at the phenomenal growth over less than 20 years, the jump through several generations of technology, and the resulting adoption of mobile phones by billions of people globally.

Even today, the wireless industry is continuing to innovate by transforming itself under the influence of a host of factors: erosion of voice traffic margins, the drive for new service delivery, shortening time to market, customers wanting more for less, competition evolving globally, and technology shifting while converging in ever-shorter timescales.

The operators are attempting to identify where to focus their efforts for long-term value creation from the assets under their control. The challenge is how to maximize the return on their investments by creating meaningful differentiation in the market. The focus is naturally on core competence and prioritization of scarce resources, both human and capital.

A Common Solution for Multiple Problems

Industry standards will help contribute to economies of scale for 802.16 solutions, so that high performance can be provided at reasonable cost. This standard will also help the industry provide solutions across multiple broadband segments:

Broadband on demand — The 802.16a wireless technology enables a service provider to provision service with speed comparable to a wired solution in a matter of days and at significantly reduced cost. It also enables instantly configurable on-demand high-speed connectivity for temporary events such as trade shows.

Underserved areas — Wireless Internet technology based on IEEE 802.16 is a natural choice for underserved rural and outlying areas with low population density.

Best-connected wireless service — The IEEE 802.16e extension to 802.16a introduces nomadic capabilities that will allow users to connect while roaming outside their home areas.

Cellular backhaul — The robust bandwidth of the 802.16 technology makes it a good choice to carry backhaul traffic for cellular base stations in a point-to-point (P-P) configuration.

Residential broadband — The gaps in cable and DSL coverage could be filled. Practical limitations prevent cable and DSL technologies from reaching many potential broadband customers.

The USP

To support a profitable business model, operators and service providers need to sustain subscriber satisfaction, a wide subscriber base, broad reach, and multiple revenue streams. WiMAX, unlike its predecessors, can fulfill all these and many more service requirements. Some of the characteristics of WiMAX that differentiate it from other existing wireless broadband solutions are explained in the following subsections (Figure 9.1).

Throughput

By using a robust modulation scheme, IEEE 802.16a delivers high throughput at long ranges with a high level of spectral efficiency that is tolerant of signal reflections. The base station can also trade throughput for range. For example, if a robust link cannot be established using 64QAM, changing to 16QAM can increase the effective range.

Scalability

The standard has been designed to scale up to hundreds or even thousands of users within one RF channel. To accommodate easy cell

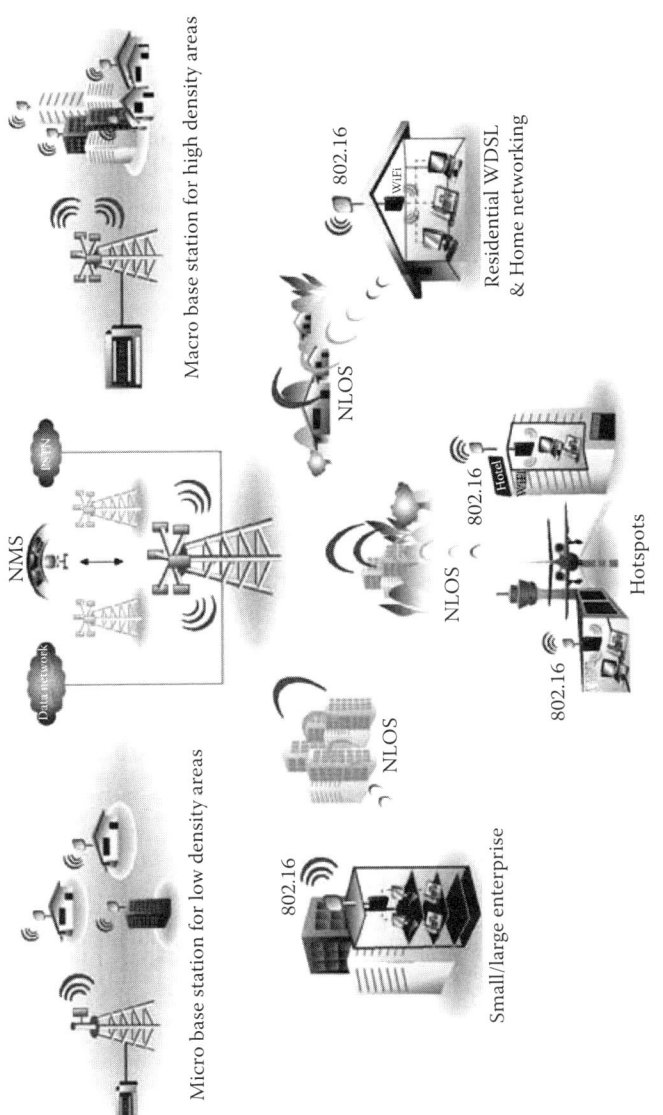

Figure 9.1 WiMAX — Broadband wireless access solution.

planning in both the licensed and license-exempt spectrum worldwide, 802.16 supports flexible channel bandwidths. Operators can reallocate spectrum through sectoring as the number of subscribers grows. An operator who is assigned 20 MHz of spectrum can divide it into two sectors of 10 MHz each.

Coverage

In addition to supporting a robust and dynamic modulation scheme, the 802.16a standard supports technologies that increase coverage, including mesh topology and smart antenna techniques.

Quality of Service (QoS)

QoS refers to the ability of the network to provide better service to selected network traffic over various technologies. The goal of QoS technologies is to provide priority (including dedicated bandwidth to control jitter and latency) that is required by some real-time and interactive traffic, while making sure that in so doing the traffic on the other paths does not fail. The standard includes QoS features that enable services which require a low-latency network, such as voice and video. The 802.16a voice service can either be VoIP or the traditional time-division-multiplexed voice.

Security

Privacy and encryption features are included in the 802.16a standard to support secure transmissions, authentication, and data encryption.

Differentiated Service Levels

The standard supports differentiated service levels. 802.16a systems can cater to a mix of subscribers having diverse service needs, i.e., a mix of business customers and residential subscribers. For example, a base station could simultaneously support more than 60 businesses with T1-level connectivity and hundreds of homes with DSL-rate connectivity.

Wider Access Scope

WiMAX adopts the orthogonal frequency division multiplexing non-line-of-sight (NLOS) propagation technology to provide broadband access to residents or enterprises for a surrounding area of more than 10 mi. In areas where wired resources are scarce and of poor quality, the advantage of WiMAX access is particularly apparent.

Flexibility

A wireless medium enables deployment of an access solution over long distances across a variety of terrains in different countries.

Standards Based

WiMAX products are based on the 802.16 standards and have to pass the consistency certification conducted by the WiMAX Forum to ensure the interconnection and interoperability of equipment of different manufacturers. The WiMAX Forum also helps in making 802.16 standards more popular, hence leading to more users, more operators, more technology providers, and faster development of new features.

Competitive Costs

WiMAX is a wireless access technology; thus, operators do not need to invest in cable installation, the construction period is short, and capacity expansion and removal is flexible and convenient. All these factors allow operators to cut capital investment, speed up capital turnover and recovery, protect investments already made, and cut business risks. This implies this following:

Lower cost of ownership
Quicker profitability
Stronger business case

Universal Acceptance: The Challenge

However, universal acceptance of 802.16 for fixed, portable, and mobile use is contingent on the industry's development in, acceptance of, and conformance to two complementary aspects of the IEEE 802.16 air interface standards work:

Development and adoption of an open and extensible end-to-end architecture framework and a specification that is independent of incumbent operators' back-end networks.

Development and adoption of means for ensuring specification-compliant and vendor-interoperable equipment to support cost-effective deployments giving users the capability to roam across networks established by different network operators.

Economics of WiMAX

WiMAX is the first widely backed wireless standard that is both technically capable and has sufficient industry support to disrupt the telecommunications landscape. It is potent enough to turn the connectivity stranglehold of incumbent telecommunications operators on its head.

WiMAX provides an economically viable broadband wireless access (BWA) technology and provides extraordinary value to service providers as well as end users. It serves new entrants as well as dominant national incumbent operators with access and backbone infrastructure:

Incumbent fixed-service operators (or ILECs)
Competitive local loop operators (or CLECs)
Wireless ISPs (WISPs)
Mobile operators

Let us first examine economic cases of existing BWA technologies that are closest to WiMAX with respect to service features. Comparative economic differentiation between Wi-Fi, WiMAX, and third-generation mobile is as described in the following text:

Economic case of Wi-Fi
    Attractive unit economics
        Customer premise equipment: per unit cost $60
        Access points or base station: $500
    Unattractive network economics
        Range: Limited — many cells, many backhaul links, for 1 sq km carpet coverage needs more than 100 access points
        Backhaul determines user experience and cost, pricing inelastic, with 1.5 Mbps typically costing $500/month, whereas 11 Mbps may cost $3000/month

Attractive services economics
    Inexpensive, sometimes free, site lease

Economic case of 3G
    Unattractive unit economics
        Spectrum cost: At 10 percent penetration, $450/subscriber
        Access points or base station: $50,000 to $100,000
    Comparatively attractive network economics
        Range: Licensed spectrum permits large cells
        Base station, backhaul costs amortized over many users
    Unattractive services economics
        Very expensive, low data rate, designed for voice

Economic case of WiMAX
    Attractive unit economics
        Customer premise equipment: Per unit expected cost of a
        WiMAX CPE would be as follows:
           – About $230 in 2005
           – About $100 in 2008
        Spectrum cost: Free license-exempt also available, low for
        licensed spectrum
        Access points or base station: $500
    Attractive network economics
        Range: Large cells
        Base station, backhaul costs amortized over many users as
        well as over 1000 subscribers that one base station and
        backhaul can service
    Attractive services economics
        Differential services provision: Can cater to the needs of a
        wide range of customers without requiring further invest-
        ment on infrastructure

WiMAX Cost Structure

The ability to provide cost-effective, affordable wireless bandwidth
(almost) everywhere is one of the key success factors for future wireless
systems. As the success of the Internet is largely attributed to the fact
that it is virtually free of (incremental) charges, it is generally perceived
that wireless data communications has to provide services in a similar
way.

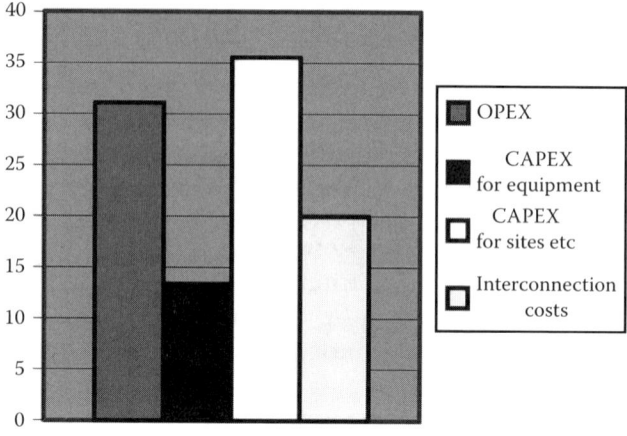

Figure 9.2 WiMAX operator infrastructure cost.

The challenge of providing flat-rate, wireless access at the cost of fixed Internet access is indeed difficult. The conventional cellular concept does not scale in bandwidth in an economical sense. Cellular systems include both the radio access network and the core network components, which have different cost and capacity performance. The more decentralized WLANs have a slightly shifted radio versus core network performance relation owing to short range and high access capacity.

Capital Expenses (CAPEXs)

These are costs related to investment in equipment and the costs for the design and implementation of the network infrastructure; site acquisition, civil works, power, antenna system, and transmission. The equipment includes the base stations, the radio controllers, and all the core network equipment. An example of CAPEX and the relations between different types of implementation costs are shown in Figure 9.2.

CAPEX components include the following:

■ Base stations
■ Site preparation
■ Service platforms
■ Spectrum

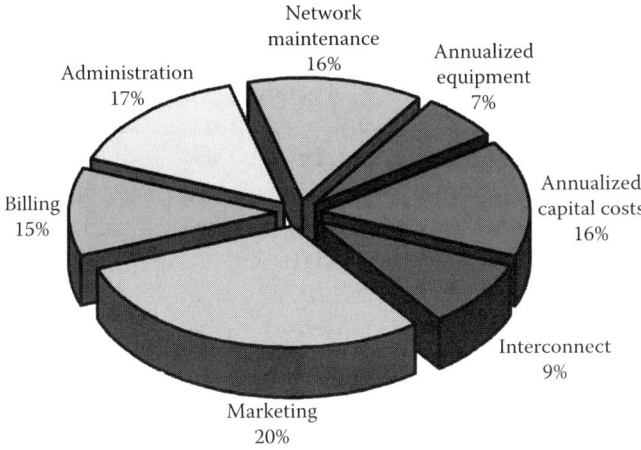

Figure 9.3   OPEX component.

Operational Expenses (OPEXs)

These are made up of three different kinds of costs:

- Customer driven — costs to attract customers, terminal subsidies, and dealer commissions
- Revenue driven — costs to get a subscriber to use the services and network or costs related to the traffic generated, service development, marketing staff, sales promotion, and interconnection
- Network driven — costs associated with the operation of the network, transmission, site rentals, operation, and maintenance

The key factors are related to customer acquisition, marketing, customer care, and interconnection. The fraction of OPEX to the overall cost changes over time; in the mature phases, the OPEX is a very vital factor. However, an estimate indicates that the network-related OPEXs are roughly 25 to 28 percent of the total costs for the full life cycle (Figure 9.3).

OPEX components include the following:

- Site leases
- Backhaul
- Network maintenance
- Customer acquisition

WiMAX Benefits

Value to Government and Society

Today, political leaders at all levels of government are working to strengthen economic development, bridge the digital divide, streamline the delivery of government services, and improve the quality of citizens' lives within their communities. To accomplish these goals, local officials are embracing a vision for digital cities, a term used to describe communities in which access technology such as WiMAX will be applied to make universal broadband access a reality and hence promote economic development and community enhancement. Specifically, this will benefit society as follows:

- Broadband telecommunication for businesses, residents, and government agencies will be universally available and affordably priced; hence, its positive impact on economic development and community enhancement.
- Solutions will be deployed to create a more efficient and responsive government while easing citizen-to-government interaction in areas such as public safety, transportation, education, E-government, healthcare, and public works.
- A formal process for cooperation between local governments and private technology and telecommunications companies means more effective technologies will emerge with these segments as a target.
- More technology investment and programs will bring technology products, services, and training to lower-income or disadvantaged areas of the community, helping bridge the digital divide.

Value to Consumers

Although market demand is not clear, technology development is driving the value for customers currently getting DSL as well as for those who do not. Existing DSL customers get far more features, including new applications and flexibility, whereas prospective customers not having DSL access can hope to get connected in a broad way. Some key benefits for customers are as follows:

- More broadband access choices, especially in areas where there are gaps, such as worldwide urban centers in which building access is difficult, suburban areas where the subscriber is too far from the central office, and rural and low population density areas where infrastructure is poor.

■ Easy and low-cost method to get connected for the billions who do not even have a basic telephone line (let alone broadband Internet).

■ More choices for broadband access will create competition, which will result in lower monthly subscription prices.

■ Payment for actual usage, and the possibility of differential service levels make optimum utility possible because service variables such as quality, speed, etc., can be selected depending upon the user need.

■ More applications and flexibility are expected later with the mobile version of WiMAX. Mobile WiMAX might bring users more potential added value than what they would get by simply replacing what they have today, e.g., increased mobility, the same provider at home and on the move, and VoIP/Skype on a PDA.

Value to Component and Equipment Makers

WiMAX promises many strategic opportunities for component and equipment makers, not just as a backhaul solution for Wi-Fi, delivering additional bandwidth to hot spots, but potentially for 3G networks too. WiMAX may also become a viable DSL or cable broadband replacement technology for consumers and may even offer nomadic or portable wireless Internet access for consumers and enterprise users. WiMAX will be an important mobile networking technology following the ratification of the 802.16e standard and the availability of WiMAX clients' devices in the year 2007–2008. Operators could also use it to carry VoIP services. The following are the implications for component and equipment makers:

■ The steady growth of outdoor wireless equipment now and indoor wireless equipment later.

■ A common platform opens the door for volume component suppliers, which drives down the cost of equipment and also creates a volume opportunity for silicon suppliers.

■ More rapid innovation because there exists a standards-based, stable platform on which to add new capabilities. A common platform allows faster innovation and accelerates price/performance improvements unachievable by proprietary approaches.

■ The amount of risk is reduced because of the economies of scale enabled by the standard. No longer does one need to develop every piece of the end-to-end solution.

Value to Service Providers and Network Operators

WiMAX can give service providers and network operators another cost-effective way to offer new high-value services such as multimedia to their subscribers. With the potential to deliver high data rates along with mobility, it can support the sophisticated lifestyle services that are increasingly in demand among consumers, along with the feature-rich voice and data services that enterprise customers require. Because it is an IP-based solution, it can be integrated with both wireline and 3G mobile networks. This versatility opens up cost-effective new opportunities for extending bandwidth to customers in a wide range of locations and for delivering new revenue-generating services such as wireless VoIP and video streaming.

Other benefits WiMAX can offer operators are as follows:

■ A common platform that drives down the cost of equipment and accelerates price/performance improvements unachievable with proprietary approaches.

■ Revenue generation by filling broadband access gaps, provision of services providing true broadband speeds, delivering >1 Mbps per user.

■ NLOS operations providing strong multi-path protection (indoor self-install).

■ High link budget enabling higher than 150 to 160 dB of link budget, high number of simultaneous sessions offering hundreds of simultaneous sessions per channel.

■ Speedy provision of T1/E1 level and on-demand high-margin broadband services.

■ Reduction of the risk associated with deployment as scalability allows investment to accommodate demand growth; also, equipment will be less expensive because of economies of scale.

■ Vendor independence as base stations will interoperate with multiple vendors' CPEs.

The Importance of Scalability and Flexibility

A scalable network provides an economical means of expanding an existing network to expeditiously meet future demands with minimal interruption in service availability caused by the expansion process.

The global demand for data services has grown at a remarkable rate in recent years. The increase in demand is likely to grow at an even faster pace in the future because of advances in multimedia distribution services. Network scalability thus becomes an important consideration for both equipment manufacturers and service providers. The overall system capacity has to be made expandable in terms of the number of subscribers supported, data rate, and geographic coverage.

There are many factors that influence the scalability of a network. Furthermore, the persistent demand for enhancement in data services is becoming an important driving force for network expansion and deployment so that the network is capable of supporting new services when they become available. At the same time, network capacity must keep up with demand.

The 802.16 standard is scalable. Imagine hundreds of hot spot users at a five-day conference trying to access the network. Accessing the local network would not be a problem as 802.11 has plenty of bandwidth within the LAN. But what if those users want to simultaneously access the Internet or hook up to their corporate network via a virtual private network? The hotel might have a single T1 connection for servicing its typical broadband connectivity use; however, for those five days, it needs a lot more bandwidth. With wireless broadband access, it is easy to ramp up service at a location for a short period of time — something wired broadband access service providers currently do not do.

It is believed that the 802.16 standard will also provide an important flexibility advantage to new businesses or businesses that move their operations frequently, such as a news van of a broadcast company, a construction or engineering company with project offices at each site, and field offices of disaster management teams. These establishments can get connected at the new location almost instantly as wireless broadband access can be quickly and easily set up at new and temporary sites.

Coexistence

WiMAX and Wi-Fi will most likely coexist with one another. Wi-Fi is great for smaller or indoor wireless networks. In contrast, the WiMAX standard was designed specifically for deployment in outdoor environments, in that it has provisions at the physical layer for optimizing outdoor conditions.

WiMAX may very well become the premier outdoor wireless network access of choice, with Wi-Fi remaining as the prevalent indoor wireless network. Together, the two will be able to provide widespread wireless coverage across both environments. Mesh networking solves connectivity challenges as well. Mesh is good for extending Wi-Fi by navigating around physical obstructions that are blocking Wi-Fi signals or by creating small, self-contained networks. In summary, all three technologies serve a purpose when extending wireless connectivity.

Convergence: The Future of Communication

Traditionally, communications media were separate and their services were distinct. Broadcasting, voice telephony, and online computer services were different and operated on different platforms. Convergence is the combination of all these different media into one operating platform. An example of traditional convergence is the combination of the personal computer and Internet technology.

The topology of computer networks has changed considerably, paralleling changes in the networking equipment and better use of increasingly powerful computer platforms, both in centralized IT and on the desktop. As the computer networks have evolved, they have come to resemble telecommunications networks.

Network convergence has ushered in an era of change for the telecommunications industry globally. Increasingly, voice, data, and video are getting deployed over a common network. The forces propelling more and more telecom players to embrace this change are the inherent benefits convergence brings, as well as the promised benefits from next-generation networks (NGNs) in the future. Convergence is seen as initial path to NGNs and is a significant issue (Figure 9.4).

In some environments, data, voice, and video traffic already travel on a single set of wires in a single network. In the future, this convergence of technologies will become commonplace, because for institutions to remain competitive, they must stay as close to the leading edge as possible with converged technologies.

Unprecedented change, extraordinary technological advances, and insatiable expectations and demands have combined to make this the most challenging period ever in the constantly evolving information

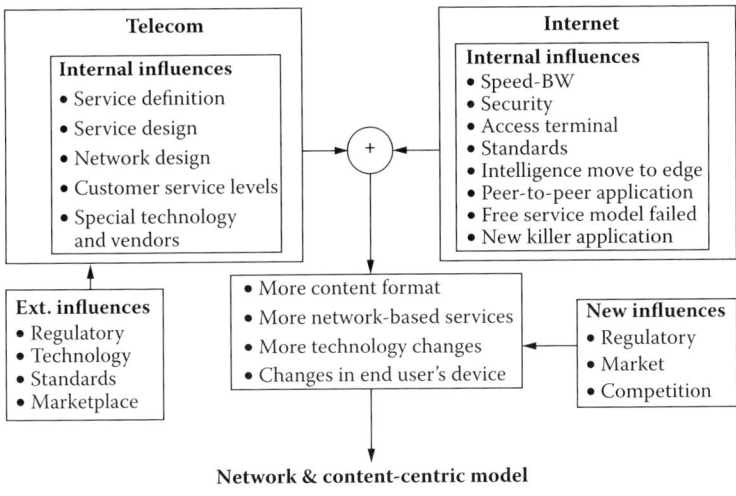

Figure 9.4 Network- and content-centric model.

communications technology. Also worth mentioning is that computer-telephony integration (CTI), though a key part of the path to maximizing the potential of integrating the computing and telecommunications environments, differs in concept from convergence. The evolution of switching the LAN has taken us beyond the CTI stage to a point where data networks, video networks, and telephone networks can be seamlessly integrated.

In this changing business environment, operators explore different ways to find new revenue streams, reduce operating costs, and provide solutions that create "stickiness" and reduce churn. The successful operator will provide a multitude of new services. Many of them will be available through both mobile and fixed access. Others will represent a combination of TV, Internet, and telephony — all of them being converged services.

Technologies that enable converged services exist. IP and the Internet paradigm are being introduced in all areas of communication. Rapid development of radio technology leading to increased bit rates and support for mobility enables true converged services — the same end-user service can be reached by both mobile and fixed access via the same user interface.

Operators that adapt their strategic business plan, considering the changing environment, with an early introduction of converged services will gain a competitive edge. Furthermore, the introduction of layered architecture will improve efficiency, flexibility, and enable a smooth introduction of IP multimedia subsystem (IMS), a cornerstone of efficient converged service offerings.

This convergence of technologies has the ability to offer sophisticated multimedia applications to the end user, thereby enhancing the entire service experience. The promise of a converged IP network has become a technological reality. Telecommunications managers and directors are responding to a commitment from technology planners to create a converged voice, video, and data network over IP. Convergence over IP is expected to create a single, economical, and pervasive solution to meeting the information and communication needs of today's knowledge workers.

Implications

Convergence is a dynamic interaction of end users, networks (wireline and wireless), and service providers, enabled by technology supporting personal empowerment at work, at home, and on the move.

Specifically, convergence refers to three phenomena: the convergence of the wireless and wireline industries to provide increasingly equivalent, highly portable service to customers regardless of their means of access, the convergence of the cable and telco industries to provide a common wireline capability, and the unconditional globalization of the entire communications industry.

Convergence is the merger of telecom, data processing, and imaging technologies. This convergence is ushering in a new epoch of multimedia, in which voice, data, and images are combined to render services to the users. This combination provides a convergence of data processing, images, and audio services. Recent examples of new, convergent services include the following:

- Internet services delivered to TV sets via systems such as Web TV
- E-mail and World Wide Web access via mobile phones
- Using the Internet for voice telephony

Today, when customers stress on convergence, they eventually expect a common set of features regardless of the access method or termination, or the number of network operators they use. This customer requirement is having a dramatic effect on network operators. Distinct categories, such as wireless and wireline, access and switching, testing and network management, will merely become different sides of the same coin.

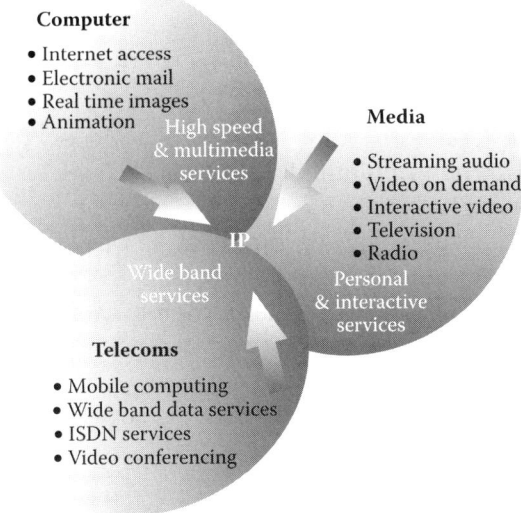

Figure 9.5   IP — the path to convergence.

Moving toward IP

Convergence will be driven by the simple fact that the telecom network is today the access network par excellence to the Internet for the general public. Increase in data traffic within the telecom network is driving operators toward a datacentric network; therefore, they naturally seek to transport both services over a common transport coming from the data world, that is, packet (or IP) networks (Figure 9.5).

Convergence of wireline and wireless access technologies is made possible by the adoption and deployment of the IP Multimedia Subsystem (IMS) — an industry-standard 3GPP/3GPP2 network-intelligent architecture that is appealing because of its simplicity and structure. Based on the application, session/call management, and transport layers, IMS provides the framework for service providers to offer new revenue-generating, lifestyle-enhancing services for consumers and supports cost-effective business-critical applications for enterprises over both fixed and wireless access methods.

IMS will help service providers integrate WiMAX technology, which has a protocol-independent core (ATM, IP, Ethernet, etc.) in a way that complements their 3G mobile high-speed data and wireline networks, so they can deliver seamless communications services to subscribers. As an example, a cellular service provider can integrate WiMAX with an existing or planned mobile wireless solution to do the following:

- Take advantage of an existing subscriber management system
- Benefit from already installed network management capabilities
- Utilize a cellular deployment infrastructure already in place, such as existing cell towers
- Provide common and additional services to users' homes, including new multimedia services

The trend is clear — the IP paradigm is, or will be, used in almost all areas of communication. A common IP-based network enables a multitude of common functions and, therefore, reduces planning and operation costs. The potential savings for operators are substantial and is one of the most important drivers of network convergence. And to save the best for last, when the underlying structure is more structured and standardized, other areas have more room for variation. This means that customized services, which of course still can be convergent according to the definition, can be provided efficiently.

Moore's Law: Impact on Communication

The PC industry, unlike the communications industry, has always enjoyed an 18-month half-life trend, a corollary of Moore's law. For example, PC prices have decreased by half every 18 months, but our communication costs have remained relatively the same over this period. Typically, the half-life of communication prices has been five years. Thankfully, all of that is about to change because of convergence. The new convergence industry is taking advantage of traditional PC architectures, examples being the following:

- CPUs (e.g., Motorola PowerPC and Intel's Pentium)
- Bus (e.g., PCI, Universal Serial Bus [USB], and IEEE 1394 — Firewire)

Then the same price erosion trends will be observed, and Moore's law will apply to convergence networks as well. As routers evolve to become multiservice, and proprietary hardware migrates to PC-based hardware implementations, and as corporate networking solutions move to the commodity market, the benefits will be reaped by the end user. With the communications industry opening up to new providers, the market will become increasingly more competitive. Consumers will soon enjoy this cost cutting and the competitively aggressive market in communications hardware and software (Figure 9.6).

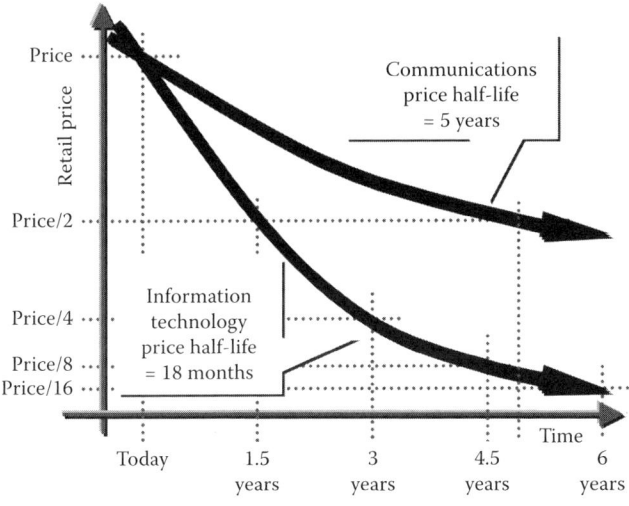

Figure 9.6   Half-life — computing and communication.

The exact modalities for the realization of the preceding convergence are far from clear today. Any detailed analysis of them would, at best, be speculative. One thing, though, is very clear — IP or the Internet world will dominate the amalgamated model.

WiMAX: The Enabler

WiMAX is slated to drive the communications industry toward unity. An ideal converged communications service is capable of providing user-portable access to high-quality voice, video, and data applications in a secured, reliable, economic, robust, and high-speed environment.

The WiMAX system based on the IEEE 802.16e standard is a very-high-throughput system and is expected to support all the features mentioned earlier. The IEEE 802.16e standard provides for mobility along with carrier-class features and reliability. Further, these standards provide for high-level data security, quality of service sufficient for voice, data, as well as video applications, high level of reliability, and guaranteed service levels.

WiMAX and Multiple Service Levels

One aspect of the existing 802.16 standards that will make it attractive to service providers and end customers alike is its provision for multiple service levels. Thus, for example, the shared data rate of up to 75 Mbps that is provided by a single base station can support the committed information rate to business customers of a guaranteed 2 Mbps (equivalent to an E1), as well as best-effort nonguaranteed 128 kbps service to residential customers.

Depending on customer demands, it should be possible for providers to offer a wide variety of standard and custom service offerings. By providing flexible services and rate structures to its customers, a WiMAX provider can appeal to a wide customer base. Operators can service the needs of different markets by a single distribution point without incurring any additional cost.

As 802.16 supports adaptive modulation, a WiMAX operator can cherry-pick subscribers, which provides excellent flexibility to the operator. WiMAX makes a strong business case for rural areas, where the distances between customers are large and they are scattered across wider areas. This allows better service spread because by using the same distribution point, an operator can provide a high data rate of 2 Mbps to business customers located close to the base station while providing 64 kbps to distant rural customers spread across a coverage area with a radius of tens of miles. This is possible as the system automatically increases effective range when necessary at the cost of decreasing throughput. High throughput can be achieved by using higher-order modulation at submaximum range, whereas lower-order modulation provides lower throughput at a higher range, from the same base station.

The modulation scheme is dynamically assigned by the base station, depending on the distance to the client, as well as weather, signal interference, and other transient factors. This flexibility further enables service providers to tailor the reach of the technology to the needs of individual distribution areas, allowing WiMAX service to be profitable in a wide variety of geographic and demographic areas.

The 802.16 standard also supports differentiated quality of service (QoS) to govern trade-offs between latency and error rate. This capability allows the technology to provide better support to different types of data transmissions. Most types of data transmission can tolerate a reasonably large degree of latency, but error rates must be tightly controlled. Real-time media such as voice and video transmissions, on the other hand, require low latency, but some degree of transmission error is acceptable. Thus, differentiated QoS permits a single data-transmission standard to handle all these different services effectively.

Marketplace Positioning

Primarily, WiMAX is envisioned as the link between Wi-Fi wireless LAN and wide area networks (the Internet, telephone system, and entertainment-programming sources). It is also seen as a way to extend broadband services to areas where CATV is not built out and where the telephone network will not support DSL. Assuming that equipment costs are competitive, WiMAX is expected to have lower initial construction costs than competing services in these areas.

WiMAX is not limited to this fixed wireless approach. Standards for mobile service have been a high priority for the 802.16 committee. It is expected that as WiMAX evolves, it will provide service directly to portable and mobile computing equipment with a bandwidth that exceeds what is possible with 802.11 WLAN.

The 802.16 committee chose the term WirelessMAN to identify the service's application as a metropolitan area network. Taking this definition to heart, several municipalities in the United States have begun development of wireless access systems that will use WiMAX as the backbone distribution network and Wi-Fi as the short-range connection to individual users.

There is significant controversy surrounding some of these systems, as some are intended to be municipal utilities. Such utilities would compete with commercial telephone, cable, and Internet companies. In most cases, the city's justification is a combination of necessary economic development and speed of development — moving ahead rapidly on new technology because of the difficult private investment climate presently being experienced in telecommunication.

Other cities have opted for a model similar to CATV, awarding one or more franchises for a metropolitan wireless network. Controversy has arisen in some cities over tax incentives and franchise terms that have been perceived as showing preference over competing service providers.

Despite the preceding target market positions, initial WiMAX operations will certainly include head-to-head competition with existing DSL, cable modem, and leased line services to small business and home office customers. This marketing effort is expected to be focused where those services are limited or, because of market size, are not priced as competitively as in large metropolitan areas. New marketing approaches may include bundling fixed service with mobile or portable hot spot access.

Market Analysis

WiMAX operators will face several classic business competition and cost issues. First, the market with the most potential users (urban areas) also has the greatest competition. If WiMAX proves to be easy to implement, a fractional market share may be sufficient to support the service.

As distance from population centers increases, the issues change from competition to market size. Suburban areas have many computer and cell phone users, but fewer business users. The cell size of a WiMAX network may be larger (and lower cost), but it will have limits because of terrain and site availability.

Rural areas may have high demand for WiMAX broadband services, but the number of potential customers will vary considerably, depending on the nature of the community. Some rural areas may present problems for high-capacity backhaul if the telecommunications infrastructure is limited.

Perhaps the most attractive market will be in distant suburban areas. Population density and cost of real estate has driven many successful people further away from city centers. This lifestyle is supported by telecommuting, which requires high-speed Internet access that can be provided by WiMAX. The number of communities with a significant population that fits this model may be limited, but it is expected to grow.

The role of WiMAX as a basic broadband access medium has already been discussed, but there are a number of marketplace uncertainties, even without WiMAX in the mix. The biggest of these is how 3G wireless services will compete with, or complement, Wi-Fi hot spot wireless access, particularly as metropolitan WiMAX and Wi-Fi networks are implemented. It is not known whether consumers will gravitate toward a single mobile broadband service or use phone-centric 3G for certain functions and computer-centric Wi-Fi for another set of uses.

A few analysts see the possibility of 3G being used mostly by individuals and Wi-Fi by business customers, but for many people those roles overlap.

After much media attention, BPL is developing very slowly and will have lower capacity than WiMAX as an alternative broadband delivery service. Also, the ability of BPL to mitigate interference to other users of the shortwave and lower-VHF spectrum has not been established. Both BPL and WiMAX are being promoted, in part, as viable options for broadband access in rural and semirural areas with no cable or DSL service.

The larger question is whether the marketplace will become more fragmented as new services are implemented or whether some services will gain some degree of domination.

Market Segmentation and Scope

The following text outlines the various market segments and their scope:

Urban
  - ■ Highest density of potential WiMAX customers
  - ■ Many multiple-tenant office and residential buildings
  - ■ Smaller WiMAX cell sizes to meet capacity requirements
  - ■ Strong competition: Driven by market size and availability of alternate access technologies

  New operators can expect the following:
  - – Lower market penetration
  - – Higher marketing and sales expenses
  Other considerations:
  - – Licensed spectrum would be desirable to minimize potential for interference
  Applications:
  - – DSL to E1 service level voice and videoconferencing

Suburban
  - ■ Moderate density of potential WiMAX customers
  - ■ Higher percentage of single-family residences
  - ■ Business parks, strip malls, etc.
  - ■ Cable and DSL may not be universally available
  - ■ Increase in WiMAX cell radius but still capacity limited with restricted spectrum assignments

  New operators can expect the following:
  - – Somewhat higher market penetration compared to urban areas
  Applications:
  - – Fixed voice and high-speed Internet

Exurban
  ■ Upscale residential neighborhoods with moderate to low household density
  ■ Fewer business establishments
  ■ High concentration of computers, cell phones, etc.
  ■ Cable and DSL not universally available
  ■ Larger WiMAX cell sizes, terrain and range probably limited
  ■ Requirements of architectural boards, environmental impact studies, etc., may add to base station site development costs
  ■ High percentage of commuters to suburban and urban areas
  ■ Expect higher market penetration for fixed broadband Internet access

Rural (small, relatively isolated cities and towns)
  ■ Distant from major metropolitan areas
  ■ Residential and small business
  ■ Very little, if any, cable or DSL (reliance is on dial-up or satellite)
  ■ High pent-up demand for Internet access
  ■ Limited competition

  New operators can expect the following:
    – Very high WiMAX market penetration and rapid adoption rate
  Other considerations:
    – High-capacity backhaul may be a challenge
  Applications:
    – Fixed voice and high-speed Internet

WiMAX Applications

WiMAX is a MAN technology that fits between wireless LANs, such as 802.11, and wireless WANs (wide-area networks), such as the cellular networks. Bandwidth generally diminishes as range increases across these classes of networks. Proponents believe that WiMAX can serve in applications such as cellular backhaul systems (in which microwave technologies dominate), backhaul systems for WI-FI hot spots and, most prominently, as residential and business broadband services.

WiMAX was developed to provide low-cost, high-quality, flexible, BWA using certified, compatible, and interoperable equipment from multiple vendors. It supports many types and flavors of wireless broadband connections, including but not limited to the following:

high-bandwidth MANs, cellular backhaul, clustered Wi-Fi hot spot backhaul, last mile broadband, cell phone replacements, and other miscellaneous applications such as automatic teller machines (ATMs), vehicular data and voice, security applications, and wireless VoIP.

Today, wherever available, these applications use expensive, proprietary methods for broadband access. In contrast, WiMAX is based on interoperability — tested systems that were built using silicon solutions based on the IEEE 802.16-2004 standard. As a result, WiMAX will reduce costs. Let us understand these applications and how WiMAX provides a compelling business case for each of them.

Metropolitan Area Networks (MANs)

Open-standard radio technologies, including 802.11, 802.16, and future standards, offer advantages to network operators, service providers, as well as users. Industrywide support and innovation are driving development of a wide array of high-performance, feature-rich, and cost-effective broadband wireless networking technologies.

Wireless Internet service providers (WISPs) have been striving for wireless technologies that make wireless metro access possible. Access to areas that are too remote, too difficult, or too expensive to reach with traditional wired infrastructures (such as fiber) require new technologies and a different approach. The three key deployment types that make up wireless metro access are backhaul, last mile, and large area coverage (referred to as hot zones).

Wireless last mile coverage typically uses the IEEE 802.11 standard with high-gain antennas, whereas hot zones use modified IEEE 802.11 equipment in a mesh deployment. Wi-Fi provides the certification for IEEE 802.11 client to access point (AP) communications. However, implementations of AP-to-AP and AP-to-service providers (that is, backhaul applications) that are typically needed for wireless last mile and hot zone coverage are still proprietary, thus providing little or no interoperability.

The IEEE 802.11 standards were designed for unwiring the LAN; hence, their use in metro access applications is facing many issues and challenges. Some of these challenges are nonstandard wireless inter-AP communication, incapability to offer economic QoS and hence voice and multimedia applications, and high cost of backhaul due to use of wires, optics, or other proprietary technologies.

Despite the challenges, wireless metro access solutions are continuously sought after because today these are more cost-effective and flexible than their wired counterparts. WISPs can offer broadband services to remote areas (such as rural towns). Local governments can provide free access for businesses or emergency services (such as police and firefighters). Educational institutions can extend learning through online collaboration between students and faculty, on and off campus. Enterprises and large private networks can communicate and monitor supply-chain activities in near real-time.

High-Speed Internet Access or Wireless DSL

WiMAX can provide high-speed wireless broadband with data speed equal to T1. Some benefits of using WiMAX for Internet access are as follows:

- Rapid deployment
- Eliminate ongoing line charges
- Service delivery to remote areas
- Affordable service available to more customers

Residential and SOHO

Today, this market segment is primarily dependent on the availability of DSL or cable. In some areas the available services may not meet customer expectations for performance or reliability, or are too expensive, or both. In many rural areas residential customers are limited to low-speed dial-up services. In developing countries there are many regions without Internet access. The analysis will show that the WiMAX technology will enable an operator to economically address this market segment and have a winning business case under a variety of demographic conditions.

Small and Medium Business

This market segment is very often underserved in areas other than the highly competitive urban environments. WiMAX technology can cost-effectively meet the requirements of small- and medium-sized businesses in low-density environments and can also provide a cost-effective alternative in urban areas competing with DSL and leased line services.

Backhaul

Cellular

In the United States, the majority of backhaul is done by leasing T1 services from incumbent wireline operators. With WiMAX technology, cellular operators will have the opportunity to reduce their dependence on backhaul facilities leased from their competitors. Outside the United States, the use of P-P microwave is more prevalent for mobile backhaul, but WiMAX, as an overlay network, can still play a role in enabling mobile operators to cost-effectively increase backhaul capacity.

This overlay approach will enable mobile operators to add the capacity required to support the wide range of new mobile services they plan to offer without the risk of disrupting existing services. In many cases, this application will be best addressed through the use of 802.16-based P-P links sharing the point-to-multi-point (P-MP) infrastructure.

Some salient points about WiMAX use as cellular backhaul are as follows:

High-capacity backhaul
- Serve multiple cell sites
- Capacity to expand and accommodate future mobile services

Lower-cost solution than traditional landline backhaul

Clustered Wi-Fi Hot Spots

Wi-Fi hot spots are being installed worldwide at a rapid pace. One of the obstacles to continued hot spot growth, however, is the availability of high-capacity, cost-effective backhaul solutions. This application can also be addressed with WiMAX technology. And with nomadic capability, WiMAX can also fill in the gaps between Wi-Fi hot spot coverage areas.

The Last Mile: Bringing Broadband to Underserved Areas

Nationwide broadband access has become a priority in many countries, both developed and underdeveloped. Still, there are a lot of gray areas when it comes to making broadband access universal. Whether developed or undeveloped, a major part of the population in most countries today is unserved and lack any type of broadband access. Let us examine broadband access scenario in both type of economies.

In most developed countries, the average broadband coverage will reach 90 percent in the coming years. Still, in most rural areas of such countries, broadband coverage will not exceed even half of that. The service gap can be categorized by two characteristics: the type of area (rural or urban) and the level of national development. In developed countries, DSL service deployment has been massive in urban and suburban deployments, whereas coverage of remote areas, that is, smaller towns and rural areas, is lagging far behind. Hurdles to overcome are the poor line quality of the installed copper base, the large distances to the central offices or cabinets, and the low population density.

In emerging countries also, the main focus of broadband deployment is on urban and suburban areas, and will remain so in the near future. Even in urban areas the low POTS penetration and the low quality of the copper pair prevent mass-scale DSL deployment and foster the need for alternate broadband technologies. Today, in most of the underdeveloped world, wired infrastructure for the delivery of residential and business broadband does not exist or is unreliable.

To extend broadband services to these underserved or unserved areas, service providers must provide new infrastructure from the ground up, which drives the price of services very high. This has been the chief cause of low density of broadband services in rural areas of the developed world; unfortunately, the situation is much worse in the underdeveloped world, where even basic telecom services are lacking.

To achieve a reasonable profit margin and acceptable timeframe, infrastructure costs must be kept under tight control. The cost factor assumes greater importance in sparsely populated areas, which are traditionally underserved by communications technology.

The second problem is sustainability and QoS. In addition to the location limitations already discussed, most of the existing technologies and solutions (especially cable services) in these places typically provide limited upstream bandwidth, which can be a substantial limitation for many business customers, depending on their specific needs. Those businesses that plan to host Web-based resources or support or plan to support a substantial remote-user base may find this limitation to be particularly significant.

In this context, WiMAX networks provide an alternative to these distribution channels, as they are independent of existing last mile infrastructure, and they provide greater upstream bandwidth than cable and DSL. These networks are also highly scalable, because providers can add additional cells to a service area at a cost that is substantially lower than that required to extend a DSL or cable network.

WiMAX, owing to the low cost and ease of deployment, along with its QoS support, longer reach, and data rates similar to DSL, is naturally positioned as a viable last mile option to offer broadband access. WiMAX as a last mile technology promises to make service delivery profitable in many regions where traditional wired technologies are impractical. A number of telecommunications carriers are considering the provision of WiMAX service in a broad range of both developed and developing countries.

In the developed countries, the densely populated and business areas are largely already served by cable and DSL. Thus, new service modalities are likely to have a hard time competing in these markets, in which the customer bases are already well served. Thus, though WiMAX will provide service differentiation in these developed segments also, it is expected to thrive in a big way largely in rural and otherwise underserved areas. Because of lack of any recognizable broadband structure in these underdeveloped areas, WiMAX will have the field to itself.

Other Applications

Automatic Teller Machines

The ability to provide ubiquitous coverage in a metropolitan area provides a tool for banks to install low-cost ATMs all across rural and suburban areas. This is a totally discounted possibility today because of the cost of satellite links and security issues with other modes of backhaul. WiMAX may bring ATMs and services kiosks to bank clients in distant suburban or rural areas. This means comfort for clients and enhanced business for banks.

Vehicular Data and Voice

WiMAX may usher in an innovation for fleet owners, logistic providers, or logistic brokers as they can find location of vehicles, their carriage capacity, and amount of loading on a real-time basis. This means better coordination for optimized carriage, unlike today when most of the carriers are plagued by low carriage on return trips. This may also help drivers and highway patrols to act speedily when facing adverse situations such as accidents or road blocks.

Video-on-Demand

Video-on-demand, one of the most hyped technologies that never took off, may get its due now. With WiMAX, a technology has been found that can make its base wider and price points better suited to the demands of customers. WiMAX allows video-on-demand to reach the masses at low cost, and hence to more people who really do need these services (unlike today when it is available to a few in city centers who have far more economical ways to get those videos). Another interesting part is that alternative videos and content related to learning, training, etc., can become a revenue-generating mechanism.

Online Gaming

If anything looks as appealing today as pornography a few years back, it is online gaming. With the emergence of this sector both in fixed and mobile forms globally, people without broadband are just waiting for technologies to make their access more pervasive and faster. WiMAX will be the technology that can provide the joysticks to rural and urban people, both at home or on the move. The wait for the child in everyone seems to be coming closer to an end.

Security Applications

Support for nomadic services and the ability to provide ubiquitous coverage in a metropolitan area provides a tool for law enforcement, fire protection, and other public safety organizations, enabling them to maintain critical communications under a variety of adverse conditions. Private networks for industrial complexes, universities, and other campus-type environments also represent a potential business opportunity for WiMAX.

Wireless video surveillance is a cost-effective, flexible, and reliable tool for monitoring traffic, key roads, bridges, dams, offshore oil and gas, military installations, perimeters, borders, and many more critical locations. Wireless video surveillance can also be used for special events, as deploying video surveillance with WiMAX as backhaul is easy and less time consuming.

Wireless VoIP

Although VoIP has been around for years, it has not been a viable alternative for most applications owing to technology constraints. Recent technology advancements have dramatically improved quality, and now VoIP service providers are positioned to offer an affordable alternative to traditional circuit-switched voice services for both businesses and consumers.

VoIP services differ from traditional voice services because the voice conversation is transmitted over a proprietary broadband network or the public Internet. This allows VoIP providers to bypass the expensive public switched telephone network (PSTN) and use a single broadband connection to transmit both voice and data. This not only reduces costs for voice providers that can be passed onto customers, but enables corporate telecom providers to layer features such as unified messaging and Web-based call control through the convergence of voice and data.

Wireless VoIP is a simple and cost-effective service that allows a subscriber to use VoIP services while on the move. This is possible because of WiMAX, which can provide carrier-grade connectivity while being wireless. It brings together the economy and benefits of VoIP and the flexibility of wireless technology (Figure 9.7).

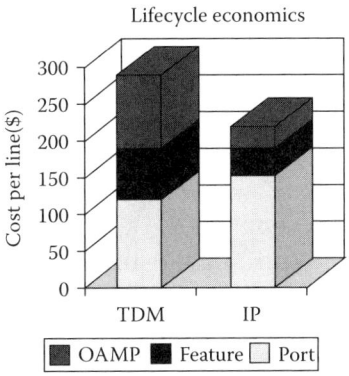

Figure 9.7   VoIP.

Multimedia Communication

IP-based wireless broadband technology can play an important role in delivering the multimedia communication, information, and entertainment that subscribers are demanding, with convenient access, anytime, anywhere. Video chats and Videoconferencing are two such services, but with different quality and different features.

Sensor Networks

Most mesh network applications, especially in the commercial sector, focus on traditional PC-based computing. However, researchers are also interested in using mesh network technologies to create networks of autonomous sensors, that is, small devices that can be installed in a variety of locations to provide readings on temperature, air quality, and other parameters.

By incorporating a wireless chipset with mesh networking software, these sensors can become network aware. After they are installed and powered on, the sensors can join a mesh network and make their data accessible to others on the network. In many situations, both in buildings and outdoors, installing small mesh-enabled sensors in many locations will be far preferable to setting up network cabling to connect the sensors or (worse) manually collecting data from the sensors.

Telematics and Telemetry

Telematics, the combination of telecommunication and computing, is predicted to be the next growth area in automotive electronics.

The use of automotive telematics in E-vehicles that have audio, e-mail and Web browsing, DVD, digital TV, and radio, as well as route guidance and traffic avoidance information, etc., is predicted to grow to more than 11 million subscribers by 2004 in the United States.

Although a related technology, such predictions have not been made for telemetry. The Formula 1 (F1) sport utilizes telemetry to beam data related to the engine and chassis to computers in the pit garage so that engineers can monitor that car's behavior. Bidirectional telemetry, from car-to-pit and pit-to-car, was allowed for a short period a few years ago. Bidirectional telemetry enables teams to alter settings on the governing electronic control unit by radio signal, and this can mean the difference between victory and defeat. However, car-to-pit telemetry is currently banned.

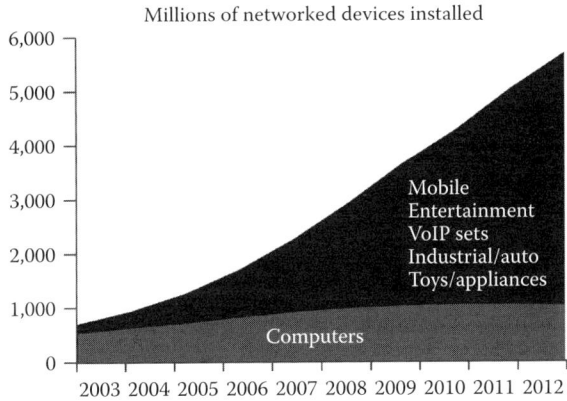

Figure 9.8 **Networked-devices projection — number of devices.**

Miscellaneous Applications

Some other possible disruptive applications of WiMAX are as follows:

- Remote monitoring of patients' vital signs in healthcare facilities to provide continuous information and immediate response in the event of a patient crisis
- Mobile transmission of maps, floor layouts, and architectural drawings to assist firefighters and other response personnel in the rescue of individuals involved in emergency situations
- Real-time monitoring, alerting, and control in situations involving handling of hazardous materials
- Wireless transmission of fingerprints, photographs, warrants, and other images to and from law enforcement field personnel

WiBro: WiMAX's Sibling

WiBro was conceived as a gap-filling technology before 3G and 4G were able to provide high-speed mobile broadband services. As the technology was waiting for a pleasant surprise because WiBro, being similar to WiMAX and with added mobility, makes for an enviable wireless broadband access technology (Figure 9.10). WiBro is the Korean wireless broadband access service based on Mobile WiMAX technology. Its key features are the following:

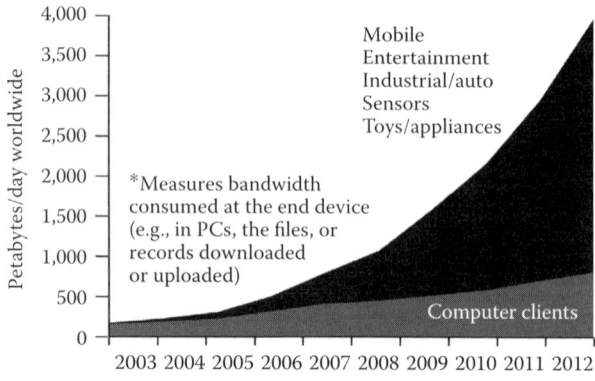

Figure 9.9   Networked-devices projection — traffic generated.

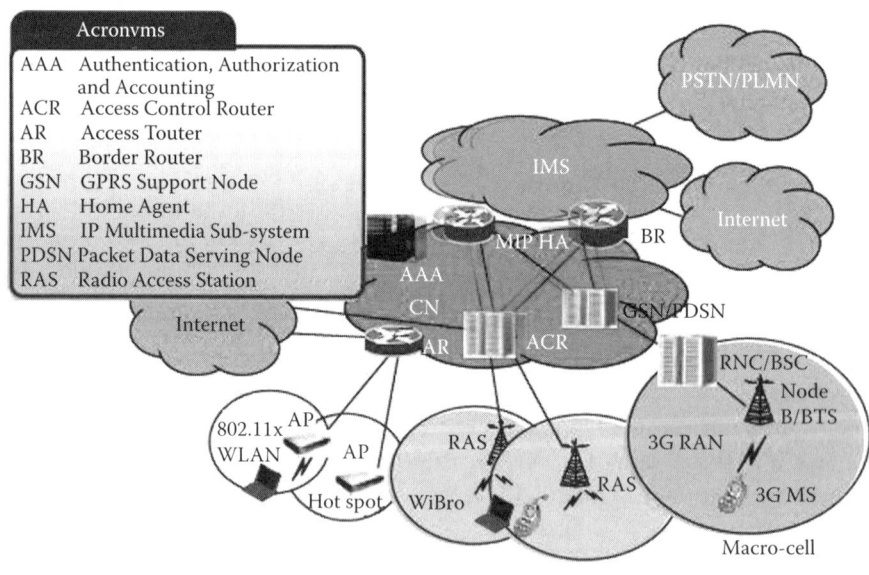

Figure 9.10   WiBro — architecture.

- Target commercial service in Korea April 2006.
- The world's first telco-grade commercial service.
- Mobile WiMAX is fully compliant with IEEE 802.16e TDD-OFDMA standards.

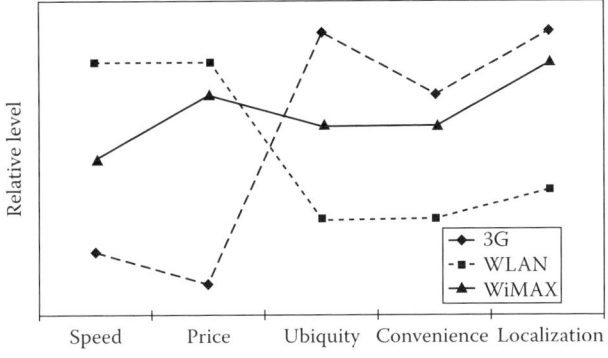

Figure 9.11   WiMAX key value comparison.

Are We Ready?

Although the telecommunications business community is not booming at this time, expectations for WiMAX are high. WiMAX appears to be a viable wireless broadband network, coming after several highly publicized earlier failures of other technologies. Through the IEEE 802.16 committee, the standards behind WiMAX have been more carefully crafted than earlier attempts at wireless broadband. In addition, the WiMAX Forum has provided a means of coordinating efforts among service providers, base stations, and customer equipment developers, chip makers, and marketing personnel.

High Expectations

The most significant issues causing concern are marketplace matters: Are there enough consumers who need the capabilities of WiMAX? Will WiMAX work with Wi-Fi as expected? Will equipment and other start-up costs meet the goals required to support deployment and ongoing operations?

Of course, these are the same questions for which all new technologies must find answers. Although stated with caution, many analysts believe WiMAX will become the primary wireless broadband system — the wireless alternative to wire, coax, or fiber broadband systems.

Deployment Process

How will a WiMAX network be implemented? The first deployments are expected to use both the unlicensed spectrum at 5.8 GHz and licensed bands at 2.5 GHz (the United States, Canada, Latin America) or 3.5 GHz (Europe, Asia, Latin America). The 5.8 GHz unlicensed band has more spectrum available (125 MHz versus 22.5 MHz at 2.5 GHz, in the United States), but allows much lower power than the licensed bands. The number of unlicensed users is unregulated and is coordinated by the operating standards. Licensed users will be protected from interference, as with all licensed services. However, licenses will be awarded by auction and will represent a sizeable investment in major markets. Unlicensed spectrum does not have this cost.

The WiMAX Forum anticipates that licensed spectrum will be used in metropolitan areas, where the increased cost is offset by the exclusivity of a license and the larger customer base. Unlicensed spectrum will see greatest use in suburban and rural areas where interference is expected to be lower. The larger capacity (bandwidth) requires fewer base stations, which should make deployment less costly in an area with fewer potential customers.

The process will begin with an analysis of the customer base — number and the distribution of locations. Next will come an analysis of the physical environment — terrain and buildings. With this information, the number of base stations can be calculated, along with their necessary locations. This process is for fixed-access deployment. Early mobile access to WiMAX will likely be limited to those places where existing base stations provide service in an opportunistic manner, with some augmentation by additional base stations when the high density of users warrants the investment.

Cell-based network engineering is well established, so the marketing process may be more important than the technical process. Attracting customers to a new service will be a significant challenge in those areas where WiMAX has the least advantage over competing broadband services.

WiMAX: Initial Phase

Initial WiMAX implementations are focused on three bands in the RF spectrum.

The 3.5 GHz band is popular outside the United States. It is the most heavily allocated band, representing the largest global BWA market. It covers 300 MHz of bandwidth from 3.3 to 3.6 GHz. The spectrum supports large-pipeline backhauling to WAN services.

The second band is the 5.8 GHz band, which has a range of 5725 to 5850 MHz. This band also is known as the upper Unlicensed National Information Infrastructure (U-NII) band. Many overlapping 5 GHz frequency bands have been earmarked for BWA growth worldwide. The World Radio Conference's 5470 to 5725 MHz band adds significant license-exempt bandwidth. Yet, most WiMAX activities are in the upper U-NII band, in which there are fewer competing services or interferences.

The third band is Multi-Channel Multi-Point Distribution Service (MMDS). Two frequency ranges reside in MMDS: the 2500 to 2690 MHz band and the 2700 to 2900 MHz band. The MMDS spectrum includes 31 channels of 6 MHz spacing in the first range. It also includes the Instructional Television Fixed Service (ITFS), which was underutilized and reallocated for BWA service in the United States. In the longer term, other bands may be useful. Examples include the two Wireless Communications Service (WCS) bands and the 2.4 GHz Industrial, Scientific, and Medical (ISM) band.

Chapter 10

Identifying the Market

Broadband access is growing fast in countries that already have well-developed fixed and mobile telecommunications networks as well as in countries with a weak telecommunications infrastructure. In the case of developed markets, the service in terms of bit rate is improving, and the tariff is gradually reducing. Although in underdeveloped economies the thrust has been to extend the reach of broadband services, the tariffs are going south while access is improving.

The major access technology today in Europe and Asia is asymmetric digital subscriber line (ADSL), or its variants, whereas in the United States, cable and digital subscriber line (DSL) are the dominant access technologies. A major global trend unfolding today is the increasing importance and prominence of wireless access technologies. Today, broadband wireless access is seen as a solution capable of filling the gaps left by wireline technologies while ensuring quality service provision in a competitive market. Because fast rollout is possible with wireless access technology, it obviously is of great interest in areas or countries with minimal or no telecommunications infrastructure.

Boom Period: Wireless Networks

Wireless LANs (WLANs) have been around for well over a decade. But until fairly recently, proprietary technologies and slow speeds have kept them largely confined to specific niches within enterprises. WLANs have long thrived, for example, on retail floors, in warehouses, and on loading docks. In these environments, tasks such as inventory checks and product code scans require broad coverage but not much bandwidth. The first IEEE WLAN standard, 802.11, was ratified in 1997 and met these requirements. Early products ran at just 1 or 2 Mbps, depending on the modulation scheme used in vendor implementations.

However, these speeds were not robust enough for mainstream business applications, particularly considering that bandwidth in WLANs is shared, not switched, which makes per-user throughput even lower. Today, however, enhanced versions of those early 802.11-based WLANs (now also known as *Wi-Fi networks*) suddenly represent one of the greatest areas of networking technology investment. Among the reasons are that the maturation of Wi-Fi technology and enterprise requirements for user mobility are finally intersecting on a fairly grand scale.

Technology contributors to the adoption of WLANs include the following:

■ The availability of products supporting higher-speed (11 and 54 Mbps) and based on IEEE 802.11 standards, making wireless networks more suited to mainstream business applications

■ Successful industry cooperation to fix known security holes unique to radio-frequency (RF) networks

■ The emergence of managementcentric WLAN architectures, which make networks easier to scale

■ The availability of automated RF tools to help networks self-adjust to environmental conditions, reducing the level of RF expertise and manual labor required by customers to install and maintain WLANs

On the enterprise demand side:

■ The convenience of wireless networking in the home has driven corporate users to demand the same flexibility at work.

■ Knowledge workers spending most of their workdays in meetings need access to corporate resources and the Internet so that they can collaborate more effectively and receive urgent communications.

■ The emergence of 802.11-based "hot spots" in public places extends an enterprise's investment in 802.11 technologies off the campus to users who are traveling.

WiMAX: Wi-Fi on Steroids

WiMAX is a much more powerful version of Wi-Fi that has been designed to have a range of 30 mi from a single well-located transmitter. Within that range, data transfer rates are anticipated to be 70 Mbps. To put that in perspective, a single WiMAX connection has the equivalent capacity of more than 500 ISDN lines, 60 T1 lines, or 7 DVD-quality video signals to each individual wireless user. A wireless metropolitan area network (MAN) based on the next generation WiMAX air interface standard is configured in much the same way as a traditional cellular network with strategically located base stations (BSs) using a point-to-multi-point architecture to deliver services over a distance up to 30 mi depending on frequency, transmit power, and receiver sensitivity.

In areas with high population densities, the range will generally be limited by capacity rather than by range because of limitations in the amount of available spectrum. The range and capabilities of the technology are equally attractive and cost-effective in a wide variety of environments. The technology was initially developed to provide wireless last mile broadband access in the MAN with performance and services comparable to or better than traditional DSL, cable, or T1/E1 leased line services.

It is likely that people increasingly will expect continued service from a broadband network when disconnected from a fixed access point.

WiMAX, UMTS, and Wi-Fi

One of the main areas of debate regarding WiMAX is where it fits in with technologies such as UMTS and Wi-Fi.

The simple answer is that it does not fit in, because the technologies differ in essential points. Let us look first at UMTS. The point is that WiMAX is built purely for data services, not for voice. It was designed predominantly for home and business users who do not have fixed-line access to broadband Internet. Voice transmission over WiMAX can only be accomplished in combination with Voice-over-IP, whereas UMTS is a technology that offers voice and multimedia services with guaranteed quality even when users are moving at high speed. UMTS is ideal for applications targeted at Internet-capable mobile phones.

WiMAX provides complementary wireless Internet access via a notebook while the user is located within a radio cell. Because WiMAX has been developed primarily for the best-effort transmission of larger data volumes at high speeds, it is a logical supplement to UMTS.

Similarly, Wi-Fi and WiMAX are complementary technologies. Because of its limited range, Wi-Fi is primarily suited for public hot spots and for office and residential use. Covering larger areas is not economically feasible. That is where WiMAX comes in, because it provides similar functionalities outside of Wi-Fi hot spots. Siemens therefore believes that future notebooks and similar data devices will contain both Wi-Fi and WiMAX technology. Most likely, operators will find a blend of technologies that helps them to optimize their cost position.

The Present Scenario

The first wave of WiMAX products were made available in the second half of 2005. These products comply with the Worldwide Interoperability for Microwave Access (WiMAX) IEEE 802.16-2004 standard. Although the overall number of WiMAX subscriber lines will at first be quite small relative to DSL or cable, the dollar value will grow to the point where even major carriers should start paying close attention.

Initially, the WiMAX Forum is focusing its profiling and certification on the Multi-Channel Multi-Point Distribution Service (MMDS), 3.5 GHz licensed, and unlicensed upper U-NII 5 GHz bands. The upper U-NII band boasts less interference, reasonable power levels, and adequate bandwidth. WiMAX-based systems deployment will begin later this year. It is not only the developed markets that can benefit from WiMAX. For emerging markets, operators are interested in using WiMAX for low-cost voice transport and delivery, which has been very difficult with proprietary solutions. In fact, growth will focus initially on markets in China, South Asia, India, and parts of South America. This growth will then move gradually into North America and Europe. Suppliers can expect steady and reliable growth. Overall, the markets without any fixed infrastructure offer the greatest opportunities for the first wave of WiMAX products.

In its initial rollout, WiMAX is likely to pose many of the same connection challenges as Wi-Fi, and users will be required to stay within the hot spot to be connected. However, a metropolitan area hot spot covering a few miles has many more applications than a Wi-Fi connection, which only reaches 300 ft.

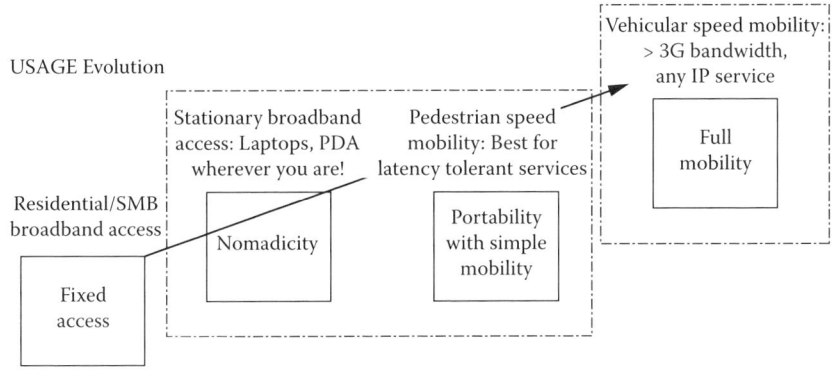

Figure 10.1 BWA usage evolution.

In 2006 as the upcoming IEEE 802.16e becomes available for wireless MANs, we will see the start of the second stage in the WiMAX evolution. WiMAX-certified chipsets embedded in laptops and other mobile devices will follow on a gradual basis. Using non-line-of-sight propagation, products such as laptops, PDAs, and cell phones will deliver services directly to the end users in a point-to-multi-point architecture. This step will lead to broadband portability and to a CPE-less business model, which makes the case even more compelling for an operator, because the user is subsidizing the model.

A more functional version of this enhanced WiMAX technology will be designed for moving vehicles. Not only is it proposed to work at highway speeds, but it also is being designed to reconnect from hot spot to hot spot, in a manner similar to cellular telephones. With the network of established cell towers already in place, some analysts are predicting that high-speed mobile wireless Internet connections could be available almost anywhere in many countries by 2008.

This technology could hypothetically equip an automobile with a browser, enabling it to receive Internet radio stations from around the world, weather maps, music, movies, television, and even home security camera video.

WiMAX is set to become the mainstream broadband wireless platform with more than 50 percent market share, used by the predicted 3.8 million broadband wireless subscribers by 2008. With its potential to replace expensive, proprietary broadband wireless — and with an evolutionary pathway already established — WiMAX appears poised for success.

Wireless Value Chain

Although the market for wireless Internet technologies has failed to meet some bullish industry projections, it has been one of the only large growth areas within the IT and telecommunications sectors during the recent economic slump, and many strategists consider it a disruptive technology.

Although parts of the wireless Internet value chain have become saturated and are undergoing consolidation, it has produced tremendous value for consumers through rapidly decreasing costs of wireless products as well as the widespread availability of wireless Internet access. The primary winners in this value chain so far have been equipment manufacturers and end users.

The players and business models in the middle of the value chain are less mature and more fragmented. Although wireless Internet markets in the United States and Europe have become increasingly saturated from the supply side, there exist many opportunities for wireless Internet in emerging markets in Latin America, Asia, and Africa, where there is an even greater need for affordable, distributive communications technology.

WiMAX Players

The broadband wireless access value chain comprises four broad segments, beginning with chipset makers, network equipment manufacturers, network software and application providers, and network integrators and operators.

Chipset makers create the core wireless technology radios and firmware (software that enables wireless networking to work with operating systems and other hardware components). These chipsets are then integrated into wireless hardware products such as client radio cards (used for laptops and PDAs) and access points, which connect client radios to an available backbone such as DSL.

Equipment manufacturers produce the basic networking components that constitute the physical layer of any wireless network. The equipment manufacturer segment of the value chain consists of access point and client radio producers, antenna and amplifier vendors, and producers of wireless-enabled access devices.

As network equipment developed, so did the need for tools to manage, operate, and protect wireless networks, which led to several new companies entering the value chain as application and software providers. The variety of these tools, combined with the increasing demand for wireless networks in public and enterprise IT markets, led to an increasing need for packaged solutions, spawning another group of new entrants (and new divisions within consulting companies) known as system integrators.

Once a critical installed base of wireless Internet-enabled devices and users was established, new market opportunities to capitalize on and operate wireless networks as an Internet access service emerged. This led to the widespread proliferation of public and private wireless Internet-enabled networks throughout the United States, Europe, and parts of Asia Pacific.

Chipset Manufacturers

As the demand for wireless networking products has grown, manufacturers of PCs, PDAs, and even cell phones have also begun to introduce wireless chipsets into their products. This segment has seen tremendous growth in recent years and has been the primary winner within the overall wireless Internet value chain. The primary reasons for this trend are that the markets have become more competitive and a system on chip provides excellent cost advantages. Second, convergence is bringing the communications industry close to computing, which has robust growth courtesy of Moore's law (according to which semiconductor capabilities double about every 18 months while costs are driven down). This allows communications players inclined toward silicon to reap the benefits of this growth in electronics.

Overall, 802.16 chipsets' roadmap have a CPE and BS chips. A subscriber station or CPE is composed of three main elements: the PHY layer, which includes a base band, the MAC (Medium Access Control) layer, and an analog RF front end that serves as the means to place signals into a specific frequency band. Equipment vendors look to chip makers to provide complete reference designs, bill of materials, components, and software/firmware to manufacture WiMAX-certified equipment.

Broadband wireless has evolved from an obscure acronym to the next big thing thanks to the marketing machines of giants such as Intel and Fujitsu. Intel is focusing on CPE, whereas Fujitsu leads the charge for base wireless terminal (BTS) development, and are to BTS what Intel is to CPE. As for other players, the Canadian company Wavesat is targeting this market and has extensive experience developing OFDM chipsets. Whether a small company such as Wavesat can triumph in this game is a question that only time can answer.

The Present

The first 802.16-2004 compliant chip was shipped to BWA equipment manufacturers by Montreal-based Wavesat Inc. some time ago, though unconfirmed sources also reveal that Intel has also shipped compliant chips recently. Wavesat's DM256, which is currently being shipped, is a baseband IC designed by Wavesat and manufactured in France by Atmel. It is being shipped only to current Wavesat customers who issued POs varying from a few samples to several thousands of units. The chip meets all WiMAX and 802.16-2004 specifications for both the BS and subscribers systems.

Opportunities

The mobile version of 802.16 (e) was ratified at the end of 2005. Meanwhile, chipset makers are busy tuning up their development boards. Whereas OFDM is the modulation of choice for fixed applications, various modulation schemes are being pushed by different technology players for 802.16e, with OFDMA being the leading contender.

The mobile version of WiMAX will be four times more complex than the fixed version, with different power and coverage requirements. The capabilities of OFDMA are necessary to provide true mobility and portability. As successful portable and mobile deployments will require higher link budgets and the use of smart antennas. Scalable OFDMA will be the required modulation for mobility.

The question of backward integration is very critical. It is very important that operators at least be able to use the equipment purchased in 2005 or before and that it be useable in future years.

- Shipments of WiMAX chipsets will likely pass 1 million units in 2007–2008.

Equipment Providers

Equipment manufactures integrate various pieces such as chips, antennas, and other boxes to create network solutions. They are one of the biggest buyers of chipsets and deliver the equipment to operators.

The next couple of years is the period of baptism for the WiMAX equipment maker. Every major wireless equipment maker is now part of the WiMAX Forum. Standards are being ratified, and the interoperability tests among vendors have begun.

The Present

The fixed/portable broadband wireless equipment market has grown by 30 percent this year. For the first time in its history, vendors, including Airspan and Alvarion, have made modest but positive cash flows. Further, broadband wireless has made formidable progress because of the increasing influence of the WiMAX forum, membership growing to the extent that WiMAX is now synonymous with broadband wireless.

Most of the BWA/WiMAX system vendors and larger infrastructure suppliers have made some progress. During the past few months some of them have also deployed or attempted to deploy proprietary broadband wireless system with specifications similar to or same as the future WiMAX solutions. Current chipsets are custom-built for each BWA vendor, making equipment development and manufacturing both costly and time consuming, which is not helping the BWA market.

These proprietary systems will be phased out only gradually and coexist in hybrid networks with WiMAX-certified solutions for obvious reasons such as cost and availability of these solutions. Such WiMAX-certified solutions will not be implemented commercially until at least Q1 2006.

OEM relationships have become key for system vendors hoping to grab a share not only of the soon-to-be-commoditized fixed WiMAX market, but more importantly, to position themselves among large mobile operators who will continue to shop with their traditional large suppliers.

Alvarion is the market leader with 26 percent market share, and continues to beat every competitor on several important business metrics such as customer base, OEM relationships, installed base, revenues, and financial position. It may not have the highest-performance system in the market, but it is definitely well in sync with market needs.

WiMAX has the potential to be as big a hit as was wireless since Marconi, but significant milestones must be achieved during the year ahead.

Opportunities

The purpose of standardization is to reduce equipment and component costs through integration and economies of scale, which will in turn allow for mass production at lower cost. With large volumes, chipsets could sell for as little as $25 and other WiMAX components could benefit from these mass volumes as well. The cost reduction will have a great impact on the CPE, driving down the average selling price to about $100 by 2008.

BS costs are more complex because of the variety of types and scale. However, BSs are less of a factor in the economic equation for operator deployments. A notable initial benefit of WiMAX is to reduce customer confusion through a WiMAX-compliance label. However, the hype generated by the press and vendors has painted an overly optimistic picture of what WiMAX systems can actually deliver. The fact is that no system can go beyond the laws of physics; every deployment will face different challenges.

Service Providers

With the multitude of wireless networking hardware and software products on the market, entrepreneurs and larger telecom and IT consulting companies have created custom solutions to bring wireless Internet to vertical markets such as healthcare, hospitality, utilities, real estate, retail, warehousing, field service and sales, and last mile communications. In addition to these vertical markets and home and office networking, wireless Internet has also been deployed to operate public (free) and private (commercial) networks, or hot spots.

Private hot spot operators have struggled to develop a successful business model, which has caused some very ambitious operators to fail to meet expectations. This results partly from other providers (ranging from homes to restaurants with Internet connections and wireless Internet access points) offering wireless Internet access for free. There is a large population of consumers who have grown to expect wireless Internet access as part of a service offering. In this sense, wireless Internet access provides businesses such as hotel chains and corporate offices the benefits of reduced networking costs and a differentiation factor for marketing, but it may not yield increasing returns as a new revenue stream. The entire value chain has also been driven and reinforced by the wireless Internet industry and standards bodies that have promoted and tailored wireless technologies to suit the needs of key stakeholders.

The Present

Wireless broadband is becoming a necessity for many residential and business subscribers worldwide. There were close to 130 million broadband subscribers worldwide at the end of 2004, a 30 percent growth from 2003. Although DSL and cable are poised to remain the dominant technologies for access in urban and developed areas, prestandard wireless access technologies are already becoming reliable and cost-effective complements or alternatives for providing voice and data services.

Today, 1 million subscribers worldwide have some form of fixed broadband (+256 kbps bidirectional) wireless access, with service revenues in 2004 touching $1.4 billion. In developing countries, representing most of the worldwide population, the potential for BWA/WiMAX growth is most pronounced. In rural areas, governments at all levels are driving the growth of broadband wireless through continuing frequency allocation and subsidies to make the rural business case more attractive, the goal being to reduce the digital divide.

Opportunities

A tipping point that will drive increased WiMAX service demand is likely to occur because of decline in cost attributable to the effects of standardization: commoditized IC or SoCs will help drive the price equation, stimulate increased awareness and market-driven demand, and provide increased supply stability and compatibility across similar equipment profiles.

The market is also anticipating a boom period owing to guaranteed fulfillment of key expectation of service providers, which is lower CPE equipment cost, ideally in the sub-$300 range. Considering the impact of CPE subsidies on the total WiMAX business case, one can easily understand what a sub-$100 CPE or a CPE built in and integrated with laptops, PDAs, or computers can do.

The second disruption, and hence enormous rise in subscriber base, is realistically aligned to effective throughput, as appeal is higher with higher throughput. Other factors playing substantial roles in future rollouts and addition to subscription numbers are interoperability, ease of installation, or coverage.

Contrary to the common belief, the majority of service providers are excited about the prospects of mobility but believe success of mobile broadband will be driven by the development of user-friendly applications and handsets and not by the implementation of 802.16e.

Software and Application Providers

The application and software provider segment consists of providers for network security, network optimization, network management, and integration with back-end systems for accounting, billing, and customer-relationship management.

One of the caveats of wireless networks has been their vulnerability to hackers, prompting a wide range of providers to develop security applications designed specifically for wireless Internet. This is perhaps the most saturated segment within the entire value chain, with dozens of competing providers.

Other companies have recognized the need for software that makes wireless networks (and the engineers who deploy them) "smarter." Application areas include network testing and optimization tools, advanced routing and power management protocols, and network training programs.

Perhaps the most interesting feature within this segment is mesh networking. Another important area is network management software, which enables network administrators to monitor and administrate wireless networks, and other existing network management software, which enables network administrators to monitor and administrate wireless networks.

Finally, there has been a particular need to develop applications that enable wireless Internet service operators to monetize network usage and tie network management software into other enterprise and legacy systems critical to business operations.

The Present

A tipping point that will drive increased WiMAX demand is likely to occur due to effects of new applications such as gaming, or video on demand (VoD), or VoIP. To satisfy the prospective end user, profitably operators will explore the challenge of growing broadband ARPU. There will not be a single solution: faster speeds and VoIP will work for some, content and IPTV services for others.

With VoIP services continuing to show strong growth and acceleration in subscriber base in both the consumer and enterprise segments, more software and application providers will become active in these areas, and more telcos will enter this segment of VoIP.

Opportunities

It will also be interesting to see the applications driving the mobile broadband market in future, which would include mobile gaming, multimedia messaging, gambling, and other applications such as ring tones. The mobile consumer market represents the lion's share of mobile data services revenue due to gaming. Also, the content for VoD will show a lot of promise, with the possibility of E-learning or off-location training.

Many companies within this segment of the value chain are likely to be consolidated into larger IT consulting firms and system integrators because of oversaturation.

WiMAX Forum

> [The purpose of the forum is to] facilitate the deployment of broadband wireless networks based on the IEEE 802.16 standards by helping to ensure the compatibility and interoperability of broadband wireless access equipment.
>
> **WiMAX Forum**

The Worldwide Interoperability for Microwave Access (WiMAX) Forum™ is a nonprofit association formed in 2003 to promote the adoption of equipment that is compliant with IEEE 802.16. The WiMAX Forum™ is an association of industry leaders dedicated to bringing about standards compliance, and interoperability based on those standards, to the wireless broadband industry. WiMAX includes well-known industry leaders, including chip makers, system vendors, network operators, industry associations, and test laboratories.

The WiMAX Forum is driving standardization and has started the process with its testing and certification program, WiMAX Forum Certified™, scheduled to begin in late 2005. This program will provide certification for equipment that meets interoperability standards, which will be a subset of the IEEE 802.16 standards.

WiMAX Forum and Wi-Fi Alliance

The organization is following the path of the Wi-Fi Alliance as a certifying agency that bases itself on the work of an industry standards body. In the case of WiMAX, the group has developed "system profiles" of the technology that the IEEE 802.16 standard specifies. The organization's members feel that these profiles will serve the needs of the broadest segments of the potential market. Moreover, the dialogue between the

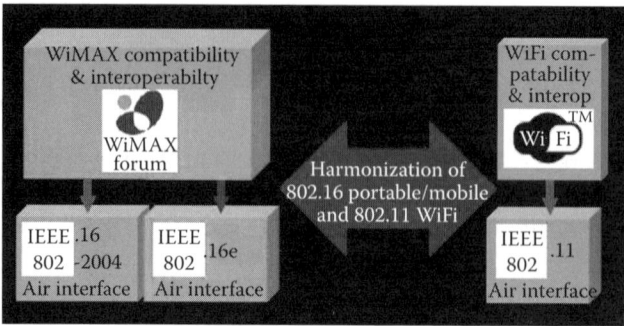

Figure 10.2 Worldwide standards and harmonization.

802.16 group and the WiMAX Forum goes both ways, because the forum feeds conformance documents back to the 802.16 group.

Although the WiMAX Forum is following the trail that the Wi-Fi Alliance blazed — in part hoping to ride the Wi-Fi momentum — the WiMAX Forum has a much tougher job. Wi-Fi is a well-defined standard in a couple of frequency bands, whereas WiMAX is not. It is probably improper to oversimplify the Wi-Fi world, but WiMAX is far broader. In fact, WiMAX will include LOS profiles in various frequency bands ranging from 10 to 66 GHz and data rates as high as 134 Mbps. Lower-than-11-MHz NLOS profiles will offer the point-to-multi-point operation that last mile broadband schemes require.

The Role of the WiMAX Forum

The nonprofit WiMAX Forum organization was formed by Intel and a number of other leading communications component and equipment companies to ensure that the interoperability issues encountered with 802.11 would not be repeated. IEEE creates standards, but it does not have a process for driving conformance, compliance, and interoperability.

The WiMAX Forum is charged with helping promote and certify the compatibility and interoperability of wireless broadband equipment. During the next year, the WiMAX Forum will develop conformance test plans, select certification laboratories, and host interoperability events for 802.16 equipment vendors. The group will also work with the European Telecommunications Standards Institute (ETSI) to align test plans for HIPERMAN, the European broadband wireless metropolitan area access standard.

Table 10.1   WiMAX Forum: Initial Profiles

Profile Name	First-Stage Profile Configuration
3.5T1	3.5 GHz, TDD, 7 MHz
3.5T2	3.5 GHz, TDD 3.5 MHz
3.5F1	3.5 GHz, FDD, 3.5 MHz
3.5F2	3.5 GHz, FDD, 7 MHz
5.8T	5.8 GHz, TDD, 10 MHz

Creation of Profiles

The IEEE 802.16 Air Interface Specification is a very capable, although complex, specification. There are allowances for a number of physical layers for different frequency bands and region-by-region frequency regulatory rules. There are features that allow an IP-centric system or an ATM-centric system, depending on the needs of customers. The specification is designed to cover applications in diverse markets from very-high-bandwidth businesses to SOHO and residential users.

An implementer currently faces a tough decision because of the wealth of options available. To address this issue, the development of system profiles was undertaken. The purpose of these system profiles is to specify which features are mandatory or optional for the various MAC or PHY layer scenarios that are most likely to arise in the deployment of real-world systems. This allows vendors addressing the same market to build systems that are interoperable while not requiring the implementation of every feature.

Creation of Test Specifications

Test specifications are necessary for the following:

To ensure that equipment and systems claiming compliance to the standard or a profile have been sufficiently tested to demonstrate that compliance.

To guarantee that equipment from multiple vendors has been tested the same way, to the same interpretation of the standard, thus increasing the interoperability of the equipment.

To enable independent conformance testing, giving further credibility to the preceding two items.

This test specification initiative is an area in which ETSI has an official process and is typically more complete than the IEEE process. ETSI follows the guidelines of the ISO/IEC 9646 series (ITU-T X.29x series). The Test Suite Structure and Test Purposes (TSS&TP) document and the Abstract Test Suite (ATC) specification, both described in ISO/IEC 9646-2 (ITU-T X.291), suit the purpose particularly well.

WiMAX test documents are as follows:

- Protocol Implementation Conformance Specification (PICS) in a tabular format
- Test Purposes and Test Suite Structure (TP and TSS)
- Radio Conformance Test Specification (RCT)
- Protocol Implementation eXtra Information for Testing (IXIT) in a tabular format

Certification

Certification is a combination of conformance- and interoperability-testing scripts based on selected profiles, with test conditions specified from the PICS document. The selection of test cases for certification is currently under development by the WiMAX Forum. Development of the certification program is one of the many activities under the auspices of the WiMAX Certification Working Group (CWG).

Certification testing is intended only for complete systems such as a base station (BS) or a subscriber station (SS), not individual solution components such as radio chips or software stacks. The introduction of BS/SS reference designs may also be considered for testing to show that the design conforms to the IEEE 802.16 specification and is interoperable with other WiMAX Forum Certified equipment, but will not obviate the requirement for a system vendor using components from the reference design to submit their product for certification testing.

For portable and mobile platforms, various vendors are expected to introduce client-based cards later that plug into a notebook or other portable platform. Such products will necessitate submission of the client-based cards with a notebook for testing similar to what has been done by a Wi-Fi certification lab. Although much work still needs to be done, as the IEEE 802.16e standard becomes more stable, the working groups within the WiMAX Forum will continue to lay down the framework for test integration and certification in migrating from IEEE 802.16d support toward the introduction of IEEE 802.16e-based products.

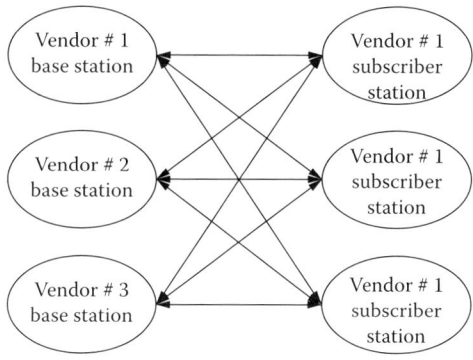

Figure 10.3 Ideal interoperability configurations.

Conformance versus Interoperability

WiMAX conformance should not be confused with interoperability. However, the combination of these two types of testing make up what is commonly referred to as *certification testing*. WiMAX conformance testing can be done by either the certification laboratory or another test laboratory and is a process in which BS and SS manufacturers will be testing their preproduction or production units to ensure that they perform in accordance with the specifications laid down in the PICS document. Based on the results of conformance testing, BS/SS vendors may choose to modify their hardware or firmware and formally resubmit these units for conformance testing. The conformance testing process may be subject to a vendor's personal interpretation of the IEEE standard, but the BS/SS units must pass all the stipulated test conditions laid down in the test plan for a specific system profile.

On the other hand, WiMAX interoperability is a multivendor (3) test process hosted by the certification laboratory to test the ability of BSs and SSs from one vendor to transmit and receive data bursts from another vendor BS or SS based on the WiMAX PICS.

The WiMAX Certification Process

First, the vendor submits BSs or SSs to the certification lab for Precertification Qualification testing, in which a subset of the WiMAX conformance and interoperability test cases is done.

These test results are used to determine if the vendor products are ready to begin the formal WiMAX conformance testing process. Upon successful completion of the conformance testing, the certification laboratory can start full interoperability testing.

However, if the vendor BS or SS fails some of the test cases, the vendor must first fix or make the necessary changes to the products and provide the upgraded BS or SS with the self-test results to the certification laboratory before additional conformance and regulatory testing can be done. If the vendor fails the interoperability testing, it must make the necessary firmware or software modifications and then resubmit the products with the self-test results for a partial conformance testing, depending on the type of failure and the required modification.

The purpose is to show service providers and end users that as WiMAX Forum Certified hardware becomes available, service providers will have the option to mix and match different BSs and SSs from various vendors in their network in their deployments. Upon successful completion of the described process flow, the WiMAX Forum would then approve and publish a vendor's product as WiMAX Forum Certified. It should be pointed out that each BS or SS must also pass regulatory testing, which is an independent parallel process to the WiMAX certification process.

Today, many solutions are customized and not always interoperable. Every piece of WiMAX Forum Certified equipment will be interoperable with other certified equipment. Additional optional features may be provided for some vendors at the risk of becoming a stand-alone system. Service providers will be able to buy equipment from more than one company and be confident that everything will work together, provided any optional features are disabled.

Market Drivers

Emergence of Standards

Standards are important for the wireless industry because they enable economies of scale that can bring down the cost of equipment, ensure interoperability, and reduce investment risk for operators. Without industrywide standards, equipment manufacturers must provide all the hardware and software building blocks and platforms for themselves, including the fundamental silicon, the SS, the BS, and the network management software that is used to provision services and remotely manage the subscriber station.

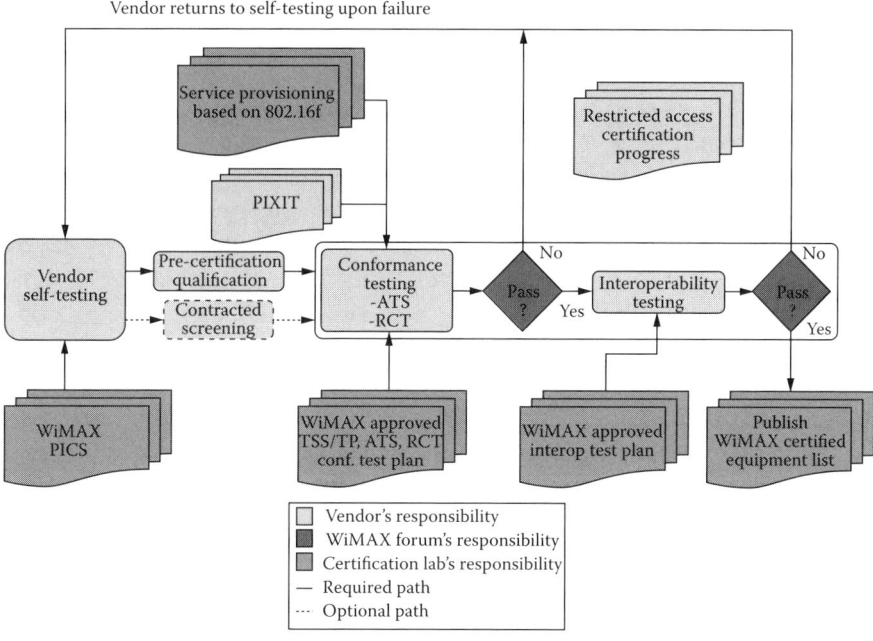

Figure 10.4   WiMAX Forum certification process.

With the 802.16 standard in place, suppliers can amortize their research and development costs over a much higher product volume. For example, a volume silicon supplier can supply the same standard component to many equipment makers at a far lower cost than would be possible if the device manufacturers were required to develop proprietary silicon for use only by their equipment.

Standards also specify minimum performance criteria for equipment, enabling a common broadband wireless access baseline platform that equipment manufacturers can use as the foundation for ongoing innovations and faster time to market. With its broad industry support, the 802.16 standard allows device manufacturers and solutions vendors to do what they do best, achieving overall price/performance improvements and opening mass-market opportunities that cannot be matched by proprietary approaches.

The Backing of Intel

The evolving world of WiMAX gained a new aura of respectability when silicon kingmaker Intel threw its weight behind the prestandard broadband wireless technology in 2003 and its investor arm, Intel Capital, pledged $150 million to help nurse the fledgling sector. It is Intel that envisaged the concept of very-low-cost WiMAX CPE evolving as quickly as possible to jump-start a mass market, hence further lowering the cost.

The Intel game plan may look as though it is aimed at reducing the dominance of a few players by pushing interpretability and standardization. However, the fact is that even though multiple players will operate at the equipment level, Intel will dominate at the chip level. Considering the explosive growth of mobiles especially in comparison with PCs, and the fact that Intel has very little or no presence in the former area, the aim is to become a key player in one of the hottest future markets.

The process of price competition and commoditization, which follows standardization, is seen as a game that Intel can play well and win. Intel's success in PC chip segment is sufficient evidence to support this theory. Ironically, this opportunity is also Intel's greatest ever challenge — the change in the nature of its PC heartland, the prospect of competition from IBM's Cell chip.

It now seems Intel's first WiMAX chipset, Rosedale, may be another on the list. Rosedale will start the WiMAX clock in terms of building market momentum, and it will reach the pinnacle of market leadership with need for mobility. Volume availability of mobile-enabled WiMAX in its Centrino chipset is expected by 2007.

Breaking the Duopoly

The global telecom market is marching inexorably toward a telco/cable duopoly. Some competitive players are showing a willingness to fight amid the growing despair of others. With cable giants growing their voice-over-IP efforts and large telcos pursuing triple-play services and megamergers, all competitive hopefuls are feeling the squeeze.

Against the backdrop of expectations that this situation is going to improve with other players emerging on the horizon, WiMAX is taking the battle to the enemy's camp. The promise of delivering DSL- and cable-level quality more economically makes it an interesting trend.

Further, the flexibility offered by WiMAX is phenomenal as a totally new operator can create a wide and dense telecom network at the fraction of the cost of the incumbent, whereas incumbent weather cellcos or telcos can leverage their existing investment by making use of WiMAX to reach a wider customer base. In which direction the balance will tilt is difficult to suggest at this early stage, but one thing is certain, WiMAX will break the duopoly, making competition multi-dimensional, and subscribers will be the biggest gainers.

Impact on Municipal Endeavors

For fixed networks, replacement has always mandated a rather straight-forward return on investment aligned closely with institutional linear thinking. The linear network is now being replaced by wireless networks that attain returns in a more holistic manner because savings can be attained from sources previously not included in a budgetary analysis. Not only will the costs for installations, moves, and changes show instantly measurable differences, but the improved accessibility to both information and collaboration alone will decrease nonproductive time. Wireless metro networks offer the greatest potential for immediate cost savings, the need for trenching, cable runs, and recurring line costs being eliminated. Imagine a major city without trenchers in the streets.

More and more municipalities, local governments, communities, and self-help groups are looking toward WiMAX as a tool to empower mobile knowledge citizens, leading ultimately to all-round regional development. What this means to WiMAX industries is huge potential markets provided they deliver metropolitan networks with the promised features.

Impact on Homeland Security

WiMAX can change the way information is shared across the battlefield and greatly influence how battles are fought and won. WiMAX, being quickly deployable and redundantly backed up, can give the soldier a tremendous advantage on the battlefield. A rugged version of WiMAX can be deployed quickly and moved about the battlefield expeditiously as needed. WiMAX will boost netcentric warfare as it can fulfill the expectations of the "digital battlefield," enabling the display of troop and vehicle locations to commanders and personnel taking part in operations.

High-quality audio and video from unmanned aerial vehicles (UAVs) can be made available to the troops who need the information. Battlefield commanders can make more informed decisions based on more accurate information from the front. This technology also promises to improve communications across the battlefield.

WiMAX can also help in other security-related applications such as surveillance of key installations across the globe from one command and control, and in case of any eventuality, decision making can be in real-time. WiMAX technology can fit in many defense projects that require agile, flexible, and high-bandwidth operations.

Market Challenges

Major challenges to overcome in deploying a WiMAX solution are the following.

RF Interference

An interfering RF source disrupts a transmission and decreases performance by making it difficult for a receiving station to interpret a signal. Forms of RF interference frequently encountered are multi-path interference and attenuation. Multi-path interference is caused by signals reflected off objects, causing reception distortion. Attenuation occurs when an RF signal passes through a solid object, such as a tree, reducing the strength of the signal and subsequently its range. Overlapping interference from an adjacent BS can generate random noise. License-exempt solutions have to contend with more interference than licensed solutions, including intranetwork interference caused by the service provider's own equipment operating in close proximity, and external network interference. Licensed solutions must only contend with internetwork interference. For license-exempt solutions, RF interference is a more serious issue in networks with centralized control than in a shared network because the BS coordinates all traffic and bandwidth allocation.

Infrastructure Placement

Infrastructure location refers to the physical location of infrastructure elements. Infrastructure placement can be an issue for both licensed and license-exempt solutions. However, infrastructure placement presents some special considerations for license-exempt solutions. Service providers are quickly deploying solutions in specific areas to stake out territory with high subscriber density and spectrum efficiency. Such areas include higher ground, densely populated or population growth areas, and areas with a less crowded RF spectrum. In addition, the physical structure that houses or supports the BS must be RF compatible. A metal farm silo, for example, may distort signals, or a tree swaying in the wind may change signal strength.

Government Regulations

Before looking at the competitive marketplace, it is important to understand the limitations that businesses have to live with.

International institutions and national governments regularly publish announcements about the allocation of spectrum portions to certain applications and service providers. Beyond the simple sharing of a limited spectrum, however, the challenges facing wireless technology implementation in the developed and developing worlds involve many different actors. These include governments, regulatory agencies, local governments, and in many developing countries, incumbent telecom monopolies.

Incumbent Telecoms

Whereas developed countries have started or completed deregulation of the communications markets, many emerging economies still run their telecom networks through a single, often state-owned organization. These monopolies are the de facto sole providers of technology, transmission, and content.

Over the years, these organizations have invested heavily in wire and cable communications infrastructure and are not yet ready to support wireless initiatives, much less to open the way to what may be perceived as potentially disruptive low-cost competition. This protective stance is often exacerbated by financial constraints because wire and cable infrastructure have been financed through debt. Quite often, incumbent telecoms also maintain tight control of national Internet backbone resources, making it difficult if not impossible for new wireless Internet service providers to operate.

The pressure exerted by such organizations to limit the access to new technology and service providers is, in some cases, encouraged by political restrictions on information access. In most cases, the result is a wireless market strewn with procedural hurdles that can prevent new actors from entering the field.

Most governments across the globe have decided in favor of spectrum allocation for BWA in the 2.3 GHz and 2.5 GHz bands. As this spectrum has already been allocated for WBA and other advanced wireless services in many countries, new entrants are also following suit with very few exceptions. What this mean is that this multination commonality or harmonization of spectrum allocations will increase the potential for realization of economies of scale and lower equipment costs, hence the greater chance of the technology succeeding.

In a first step toward worldwide spectrum allocation to wireless Internet applications, the ITU in its July 4, 2003, World Radio Conference communiqué indicates that it "successfully established new frequency allocations to the mobile service in the bands 5150 to 5350 MHz and 5470 to 5725 MHz for the implementation of wireless access systems including RLANs. Wireless devices that do not require individual licenses are being used to create broadband networks in homes, offices, and schools. These networks are also used in public facilities in so-called hot spots such as airports, cafes, hotels, hospitals, train stations, and conference sites to offer broadband access to the Internet ....

"The lower part of the 5 GHz spectrum will be predominantly used for indoor applications with the first 100 MHz (5150 to 5250 MHz) restricted to indoor use. The use of these frequency bands is conditional to provisions that provide for interference mitigation mechanisms and power-emission limits to avoid interference with other radio communication services operating in the same spectrum range."

Another common and widespread regulation already in place in several countries is that there are essentially no restrictions that limit the use of specific technologies in a given band. Many countries allow 3G operators to deploy non-IMT-2000 WBA technologies in their 3G spectrum as this will afford the operators greater flexibility in offering a broad range of services to meet customer needs.

In some markets a welcome step by regulators had been the availability of licenses for WBA to several potential providers. A multi-operator, competitive environment will offer maximum benefit to customers in the form of quality services and reasonable pricing. Accordingly, eligibility of providers for licenses in the WBA spectrum must be set carefully. The eligibility criteria must try to accommodate existing providers on the basis of their past records, but it is equally important to ensure that qualified new operators are not excluded from bidding and possibly winning the available WBA licenses.

QoS standards should not be established by regulatory mandate. Characteristics such as network availability and network latency are best supported by a competitive marketplace in which customers demand and operators supply feature-rich, quality services. They can support competition through the allocation of appropriate spectrum blocks for use in a given geographic or market area (e.g., islandwide), and the provision of the associated licenses that will create a multioperator environment. Technical standards for WBA services are best established by standards development organizations. Implementation of such standards should be voluntary and implemented based upon each operator's competitive and business decisions.

Licenses

Today a substantial chunk of prime lower-frequency spectrum is currently occupied by incumbent licensees that offer "broadband" services. Traditional over-the-air radio and television broadcasting in the AM/FM and VHF/UHF bands use valuable lower-frequency real estate for which alternative higher-frequency distribution media are already available (e.g., coaxial cable television and direct broadcast satellites deliver many more programming channels without using lower-frequency spectrum).

Partially relocating these services to higher-frequency channels or conversion to more efficient broadcast transmission technologies would free up additional commercial spectrum for other uses that currently lack viable alternatives, including allocating additional spectrum for dedicated unlicensed use.

There are three principal economic justifications for relying on exclusive licenses: (1) spectrum scarcity, (2) investment incentives, and (3) interference management. The first justification views spectrum as an economic good and focuses on the role of markets in allocating that good. The second justification focuses on the need to provide users and providers with appropriate incentives to invest in network equipment and services. The third justification recognizes that even when spectrum is not "scarce," it may be necessary to coordinate user behavior to allow users to share spectrum without adversely impacting or "interfering" with each other.

The rapid pace of innovation and technical uncertainty implies that licenses pose a risk of future expropriation in system value. As with overly broad intellectual property rules, exclusive licenses can allow a licensee to use artificial scarcity to extract the surplus associated with a new technology.

The Spectrum Picture

Wireless broadband is clearly at a crossroads. Convergence is taking place between the technology road maps of WiMAX/802.16 and advanced 3GPP, 3.5G-4G cellular systems. These technologies are on a collision course and will provide similar bandwidth and significant market overlap by 2010.

The evolution of spectrum availability and overall regulation will greatly impact the future of mobile broadband wireless systems.

Fixed Broadband Wireless Spectrum

3.5 GHz Band

The 3.5 GHz band is the most widely available band allocated for broadband wireless access worldwide, except for the United States (despite the recent opening at 3650 MHz). Covering 300 MHz of bandwidth, from 3.3 to 3.6 GHz and in some case up to 3.8 GHz, this band offers great potential for fixed applications, whether backhaul or last mile access.

3.5 GHz remains a band allocated mostly for fixed-only services in 77 percent of the countries surveyed. However the regulators are starting to revise their positions to allow portable services in a first step toward allowing full mobility at 3.5 GHz. Thirteen percent of countries surveyed have loosened up their requirements for fixed-only services at 3.5 GHz. Regulators recognize that the line distinguishing BWA and 3G is blurring, and these technologies may converge in the future.

5 GHz U-NII and WRC Bands

The Unlicensed National Information Infrastructure (U-NII) bands have three major frequency bands: low and mild U-NII bands (5150 to 5350 MHz) (802.11a), WRC (new) (5470 to 5725 MHz), and upper U-NII/ISM band (5725 to 5850 MHz). Wi-Fi exists in the lower and middle U-NII bands, which have demonstrated viability for BWA. Many overlapping 5 GHz frequency bands earmarked for BWA growth exist around the world. The newly allocated World Radio Conference (WRC) 5470 to 5725 MHz band adds significant license-exempt bandwidth. Most metropolitan deployments are in the upper U-NII 5725 to 5850 MHz band because there is less interference there, i.e., Wi-Fi and the outdoor power allowance are in the higher 2 to 4 W range compared to only 1 W in the lower and middle U-NII bands.

Multi-Channel Multi-Point Distribution Service

The MMDS spectrum includes 31 channels of 6 MHz spacing in the 2500 to 2690 MHz range and includes the Instructional Television Fixed Service (ITFS) in the United States. This spectrum has been significantly underutilized for its original instructional television purpose, and has been allocated for BWA service in a few countries including the United States, Brazil, Mexico, and Canada.

However, spectrum such as the following are not yet addressed by WiMAX's plans to focus system profiles to use OFDM modulation, operating in the 3.5 GHz licensed (non-U.S.), 5.8 GHz license-exempt, and 2.5 GHz licensed bands, in that order:

- License-exempt sharing of television broadcast spectrum
- 700 MHz
- 902 to 928 MHz (the United States and Canada)
- 2.40 to 2.4835 GHz
- 5.250 to 5.350 GHz (mid-UNII band)
- 5.470 to 5.725 GHz (proposed additional 255 MHz in the United States)
- 24 GHz
- 60 GHz
- 70-80-90 GHz

Future Spectrum for BWA/WiMAX

Additional bands are being considered today by different regions around the world for the deployment of WiMAX and other similar broadband wireless access services. In Japan the 4.9 to 5.0 GHz band will be used after 2007, whereas the 5.47 to 5.725 GHz band is also being considered for future use. The first one will require a license for BS deployment and will support 5, 10, and 20 MHz bandwidths, whereas the second one will possibly not require a license and would support 20 MHz bandwidths. In the United States, 700 MHz is slowly being freed by broadcasters to allow BWA services, and 450 MHz is seeing renewed interest for mobile WiMAX owing to its great propagation characteristics.

In the 3.6 to 4.2 GHz range, the following are important developments:

- The United States will finalize allocation of 3650 to 3700 MHz in the first half of 2005.
- Some manufacturers and service providers are starting to look at 3.6 to 4.2 GHz for 4G.
- The United Kingdom already has some FWA licenses in 3.6 to 3.8 GHz.
- CEPT (Europe) and France issued 3.4 to 3.8 GHz consultation in the fourth quarter of 2004.
- Malaysia issued 3.4 to 4.2 GHz consultation in 2004.

Block Sizes

The situation varies from region to region and across countries within the same region. In Europe, many blocks assigned are 20/25//28MHz/ or 14 MHz wide. Some countries such as Norway have assigned narrower blocks (2X3.5 MHz). The largest blocks we found were in Sweden, with 2X70 MHz. In Asia 10.5 MHz blocks in duplex are common (China, Hong Kong). In Latin America, most blocks assigned are in the 25 MHz range.

FDD and TDD Status

During our research talking to 50 regulators worldwide, we came up with the following results. We believe the trend among regulators will be "technology neutral" to provide the flexibility to operators to deploy the solutions they need. As for spectrum availability for WiMAX mobility, regulators recognize that the line distinguishing BWA and 3G is blurring and these technologies may converge in future. However, regulators must honor their commitment made in the 3G auctions to not allocate spectrum for 3G mobile communication services before a determined period of time around 2006–2007. Numerous regulators have employed the International Telecommunication Union (ITU) definition of "pedestrian mobility speed" for 3G technologies to differentiate between the two.

To be specific, this means that wireless broadband operators may only offer fixed or pedestrian mobile services. Operators are not allowed to provide mobile services at vehicular speeds for now. This restriction will be lifted once the 3G moratorium ends.

More liberal countries where full mobility is allowed include the United States (2.5 GHz), Canada (3.5 and 2.5 GHz), and Australia and Korea (2.3 GHz WiBro).

In most of Europe the band 2.5 to 2.69 GHz is exclusively reserved for UMTS mobile services and is therefore not available to BWA/WiMAX service providers. In other parts of the world, initiatives such as the ITU WP8F are pushing to allow interoperability bodies between UMTS and OFDM in these mobile services.

The ITU is organized into three main sectors. Each sector is broken up into study groups that carry out the majority of the technical work. All ITU guidelines are developed according to a formal process. The study groups address particular technical questions, which are technology areas that warrant further research. Once a topic has been sufficiently researched and a decision has been made about how to proceed, the group submits a formal "recommendation." This recommendation is then shared with all of the external ITU partners, such as SDOs and national governments.

Two groups within the ITU specifically engage in helping to define the next generation of mobile wireless. These two groups include the following:

■ Working Party 8F (WP8F) in section ITU-R
■ Special Study Group (SSG) "IMT 2000 and Beyond" in section ITU-T

WP8F is focused on the overall radio-system aspects of 4G, such as radio interfaces, radio-access networks (RANs), spectrum issues, service and traffic characteristics, and market estimations. The SSG "IMT-2000 and Beyond" is primarily responsible for the network or wireline aspects of future wireless systems including wireless Internet, convergence of mobile and fixed networks, mobility management, internetworking, and interoperability.

Beyond the regulatory constraints, WiMAX needs lower bands to economically deploy networks that will provide full mobility. Bands higher than 3 GHz are not suitable for mobile networks as proper coverage would require too many BSs compared to sub-1-GHz bands. The WiMAX regulatory group is working toward influencing the regulatory bodies worldwide to open up bands for WiMAX mobility. Those bands could include 700 MHz and 450 MHz. The regulatory working group is also working to create an environment to support eventual global roaming for nomadic and mobile WiMAX devices.

The WiMAX regulatory group is working toward influencing the regulatory bodies worldwide to open up bands for WiMAX mobility. Those bands could include the 700 MHz and 450 MHz bands. The regulatory working group is also working to create an environment to support eventual global roaming for nomadic and mobile WiMAX devices.

Chapter 11

Predicting the Future

Faced with the emergence of the gun, every smart early-seventeenth-century blacksmith had to ask himself three questions: How many of my customers will trade their swords for guns? How soon? And should I fight, adapt, or admit defeat? The pace of innovation at the start of the twenty-first century means that similar dilemmas now confront many industries — and none more so than telecommunications industry.

The Historical Pattern

In the last part of the twentieth century, the almost simultaneous arrival of two major innovations — mobile phones and the Internet — not only changed the face of communications, but also stimulated dramatic economic growth.

This historical pattern has been repeated in the development of every new communications network technology:

- 1840s: The telegraph
- 1870s: The telephone
- 1890s: Radio telegraphy or "wireless"
- 1920s: Radio broadcasting
- 1950s: Television broadcasting
- 1960s: Geostationary satellite communications
- 1970s: Computer communications
- 1980s: Optical communications
- 1990s: Internet and mobile communications

Lessons from History

Are we any good at predicting the future? When we look back over history at any advancement in electronic communications networks, we tend to forget about the highs and the lows, the boom-and-bust cycles, and the failed predictions about likely usage. We invariably get it wrong.

We got it wrong for telephone. Promoters struggled for three decades to identify the application that would promote its wide adoption by home owners and businesses. At first the telephone was promoted as a replacement for the telegraph, allowing businesses to send messages more easily and without an operator. Telephone promoters in the early years touted the telephone as a so-called killer new service to broadcast news, concerts, church services, weather reports, etc. Industry journals publicized inventive uses of the telephone such as sales by telephone, consulting with doctors, ordering groceries over the telephone, and listening to school lectures. No one wanted to believe that someone would buy the telephone to chat.

We got it wrong again for e-mail. The popularity of e-mail was not foreseen by the ARPANet's planners, the parents of Internet, and did not include electronic mail in the original blueprint for the network. In fact, in 1967 they had called the ability to send messages between users "not an important motivation for a network of scientific computers." The popularity of e-mail was a surprise, as the rationale for building the network had focused on providing access to computers and not to people.

We got it wrong once more for the dot.com. Dot.com mania and telecom hype ruled, with people believing ridiculous projections by people from WorldCom that Internet traffic has been doubling every 100 days during the period 1994 to 2000 without realizing that such an increase in traffic would imply an unrealistic market size either at the start or in 2000. Dot.com was going to change how we lived and worked, and it was widely forecasted that the Internet was about to take over as the sole communications medium. What this led to was massive overbuilding of capacity compared to the actual need. The result was that once high-flying telecom companies such as WorldCom and Global Crossing dramatically went bankrupt, leaving behind massive debts, many more teetered on the edge of bankruptcy. Billions of dollars have evaporated in stock market valuations.

We got it wrong once more for E–commerce. Undeterred by the so-called telecom and dot.com bubble bust, E-commerce got us back to our dreaming habits again. We thought that E-commerce is the only way to save this world from inefficient and ineffective business practices. It was repeated in innumerable conferences and keynote addresses that the alternative to becoming E-commerce enabled was death as only E-commerce would remain in the near future. With plenty of investments and reengineering, we still do not have many cases of E-commerce investments providing justifiable ROI.

So What's Coming Now?

With this track record, we should not even try, but one more time let us see where communications technology is moving.

If we look at past innovations such as the telegraph, the telephone, radio telegraphy or wireless, radio and television broadcasting, geostationary satellite communications, computer communications, optical communications, and Internet and mobile communications, a common trend can be discerned. All these technologies made human interconnectivity easier, better, and more pervasive.

All these innovations had more people (the masses) interconnected (bring together entire societies) easily at a low cost that kept dropping with every passing day, without requiring time to get acceptance for extraordinary initial investments. So "interconnected," "mainstream," "ubiquitous," and "low cost" are the keys to success, as was the case in the past.

Broadband Wireless Access (BWA): The Next Big Thing

BWA satisfies the preceding prerequisites completely:

Interconnect: BWA is capable of creating linkages that will bring people together. It can create networks, closing the gaps between markets, goods, and services. BWA will interconnect entire societies.

Mainstream: BWA is among the few innovations that can bring about a big change in society. It will impact all of society by ultimately becoming an item of mass consumption available to everybody.

Ubiquitous: BWA will be omnipresent. Technology advancements such as WiMAX, 4G, NGN (next-generation networks) will make BWA available everywhere, wherever they are needed.

Low cost: BWA is destined to become available at an unbelievably low cost. All indicators are that price will be going only one way: down, rapidly and continuously.

Prime mover: Applications add value to the utility of technology. Killer applications such as VOIP and VoD are going to drive BWA or WiMAX to become a prime technology for businesses as well as individuals.

The Failure of Generation One

Teligent spent $1.3 billion in a year building its network but had only signed up 35,500 customers by the time it filed for bankruptcy in 2001.

NextLink spent $695 million purchasing LMDS licenses in 1999 and, unfortunately, was not able to recover a substantial part of this investment.

Line-of-sight (LOS) and abnormal spectrum cost took a heavy toll on broadband wireless when it made its first appearance in the late 1990s. The requirement for expensive and disruptive truck roll and the difficulty of creating effective coverage in urban areas soon emerged as a lethal dose for first-generation BWA.

Another reason why costs outran revenue potential was the lack of standards. Proprietary equipment costs were high because kits had to be made separately for each operator, so vendors never got the scale to reduce prices and take on wireline infrastructure, especially in an immature market. Equipment makers had to create complete end-to-end solutions, so there was no potential for commodity subscriber units, nor would LOS technologies permit portability, limiting the scope of the applications and their usefulness to businesses.

Unsecured BWA

One of the caveats of wireless networks has been their vulnerability to hackers, prompting a wide range of providers to develop security applications designed specifically for wireless Internet. The well-publicized security flaws in previous BWA solutions, including 802.11, have served as the primary reason why more individuals and enterprises have not installed a WLAN-based BWA system.

The core flaws are now being fixed by the Wi-Fi Alliance, an industry trade group, with its Wireless Protected Access (WPA) specification, available in some products today, and by the 802.11 committee with 802.11i, a much more sophisticated package of functionality that covers both authentication and very powerful and secure encryption. One thing to be kept in mind, though, is that these techniques only secure the air link, the portion of the connection between a client and the infrastructure. If you want real security, consider using a virtual private network (VPN), which is exactly what you would do on your wired network.

Discussions on wireless security often highlight the fact that solving the security issues will be complex and difficult. This seems quite true when we try to improve security aspects of these flawed families of technology discussed in the preceding text for Wi-Fi or the 802.11 family of technologies.

However, new developments such as WiMAX or 802.16 are far more secured and promise security from day one.

BWA: The Present

Many analysts, thinkers, and industry gurus have made a number of general observations about trends in the BWA space.

First, proprietary physical layer choices are giving way to a small number of standards-based technologies such as 802.16. This is driven by the economics of the semiconductor chips required to implement these standards.

Second, there is a distinct trade-off to be made between maximizing bit rate and reach by using rooftop subscriber antennas, and reducing costs of deployment by designing for desktop or PC-card antennas that can be purchased and installed by the consumer. Serving PC-card devices well requires that the technology deal with user mobility as well.

Third, long-reach systems mean fewer cells, and typically wired backhaul. Shorter-reach systems require more cells and create a demand for distinct middle mile approaches, such as wireless backhaul or mesh networks. In the near term we are likely to see combination approaches, e.g., large-cell WiMAX base stations serving fixed antennas, used as a backhaul technology for 802.11 access points serving mobile users equipped with PC cards or antennas built into a laptop PC or PDA.

Fourth, newer-generation systems, or updates to existing standards, are being designed to handle VoIP and other applications requiring quality of service (QoS).

Finally, systems operating in the unlicensed spectrum allow rapid entry by competitive WISPs; however, congestion in the unlicensed band may lead to a preference for licensed spectrum, such as MMDS, for long-reach systems. In the near term, congestion in the 2.4 GHz band from 802.11b/g is driving WISPs to the UNII band at 5 GHz.

Taking Broadband to the Masses

Given that more than 150 million lines were in use in the world at the end of 2004, broadband has clearly become one of the fastest-growing mass-market telecommunications services in history. In 2004 alone, more than 50 million new lines were added, and analysts expect this growth to continue. Indeed, the estimates for 2009 are 400 million broadband subscribers.

At the end of the day, it is all about delivering low-cost BWA to the masses. Regardless of which technology comes out on top, it is the millions of people in urban, rural, and developing markets who need to be served.

Professional Users: Needing Mobile Broadband More Than Ever

Wireless broadband is deemed to be an important driver of productivity. For this reason, IT departments are eagerly awaiting the higher speeds that will enable workers to access corporate networks wherever and whenever necessary. When users leave the office, or even their desks, mobile broadband will enable them to access corporate networks and the Internet in a fast, reliable, and secure manner.

Given heavy corporate emphasis and strong government encouragement for implementation, it is natural to expect that a wide range of industries will adopt wireless broadband for their workforces. The public and private sectors each stand to benefit from increased productivity. In all likelihood, the fastest-growing, most price-tolerant and profitable segment will be made up of corporate users.

The mobile professional (defined as workers who spend 20% or more of their working hours outside of the office) population is exploding and will increase at an even faster rate, with lesser but substantial growth among work extenders and telecommuters. Many of these mobile professionals need wireless data services, though other groups also are prospects for wireless data. Professionals requiring wireless data include the following:

- Full-time and part-time teleworkers and remote workers, business travelers, and other workers who must frequently work away from their main place of business
- Workers who have mobile communications needs that are portable and event driven rather than truly mobile
- Workers whose business activities can be conducted from non-office-locations such as hotels, airports, conference centers, and restaurants
- Workers who are located in developed countries or developing countries with high mobile growth and weak wired infrastructure
- Workers who need to exchange and access data when conducting business activity outside their main place of business

Together, these groups will drive fast adoption of BWA. Data traffic coming from these wireless workers is increasing as never before.

Other than new wireless technologies that enable broadband access, there are a number of other factors leading to the increased adoption of wireless data, such as the following:

Globalization of business: Enterprises are expanding their presence beyond traditional markets, targeting new ones and reducing costs, taking advantage of the arbitrage offered by worldwide resources. With globalization, the business cycle is no longer limited to 8- or 9-hr workdays but a near 24-hr cycle. Mobile devices become crucial for businesses by enabling employees and partners to communicate seamlessly anytime and anywhere with positive effects on productivity.

Growing mobile workforce: According to IDC analysts, the number of mobile workers in the United States is expected to grow from 92 million in 2001 to 105 million by 2006, whereas in Western Europe the number is forecast to grow from 80.6 million in 2002 to 99.3 million by 2007. According to the UMTS Forum, the situation in other parts of world is not very different, with dramatic changes in professional and social patterns taking the number of mobile workers to 680 million globally.

Convenient wireless access devices and wireless data applications: Mobile phones are evolving from just making voice calls to serving as complete wireless personal information management (PIM) systems. Mobile device manufacturers are also developing devices that are hybrid phone-PDAs, such as Nokia's Communicator and Handspring's Treo. There have been increasing shipments of

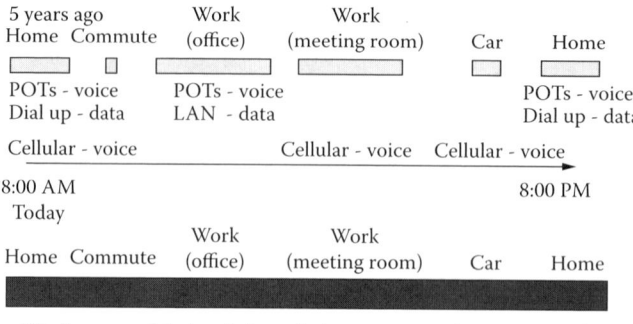

Figure 11.1 Evolution of communication for professionals.

access devices, such as laptops and PDAs, with built-in Wi-Fi, and by 2006 with WiMAX support. The higher computing power and larger screens of these devices enable a greater number of enterprise applications to be accessed. Software vendors are striving to wireless-enable common applications, such as e-mail, calendar, and office productivity tools.

Public and Residential Users: Getting More Done at Home

Advances in technology as well as saturation and more competition in professional user markets have led operators and broadband service providers to address the enterprise and consumer segments. Moreover, they have an opportunity to tap the traditional residential broadband segment in a more profitable way then ever before. From an end-user perspective, a fixed WBA service of this kind, terminated in stationary customer premises equipment (CPE), is comparable to any conventional broadband service based on DSL or cable.

Today, fixed operators face great difficulties in delivering ADSL services in rural areas where the length of the local loop is inadequate and the cost of bringing fiber closer to the customer is prohibitive. By contrast, wireless operators can successfully deliver broadband connectivity services with high bit rates over wide service areas. In rural areas, this approach might actually be the only cost-effective solution to offering broadband services. A similar approach can be used in urban and suburban areas to deliver broadband to the home or office.

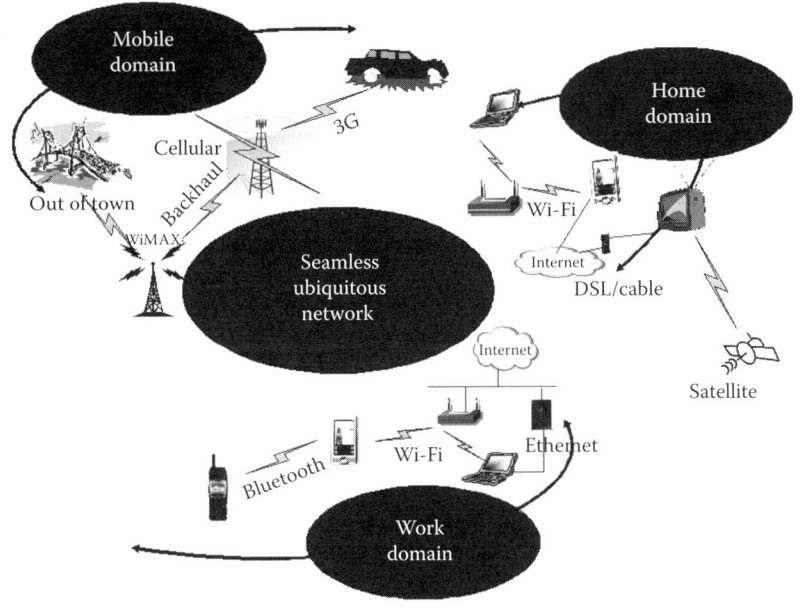

Figure 11.2   Personal broadband.

Drivers of BWA usage in the residential segment are multimedia services such as video-on-demand, voice and video peer-to-peer interaction, Voice-over-IP, and gaming. In the public segment the key is availability of affordable broadband services to communities, including rural masses, which is seen as important for a number of reasons, including attracting investment to the region, assisting regeneration of local communities, delivering E-government, supporting local economic development and community projects, promoting E-learning and remote learning, and providing online access to information.

WiMAX: The New Kid on the Block

Public broadband access via wireless is not only a boon to business travelers but is also an interesting business opportunity in itself. Broadband wireless Internet access via so-called hot spots in hotels, airports, convention centers, coffee shops, restaurants, etc., is a fast-growing trend. Hot spots provide Internet access for hire. Relatively economical to set up, all that is required to create a simple hot spot is a broadband connection and a wireless router. Many hot spots use T1 for its high bandwidth, but DSL, cable, and fixed wireless can also be used.

WiMAX can make high-speed wireless Internet services available to much larger areas than can typical Wi-Fi hot spots. WiMAX implementations can provide a wireless range of up to 31 mi (50 km), much greater than the physical distance limitations of Wi-Fi hot spots (328 ft) or DSL (15,500 ft). WiMAX can also be used to interconnect existing Wi-Fi networks.

WiMAX promises many strategic opportunities, not just as a backhaul solution for Wi-Fi delivering additional bandwidth to hot spots, but potentially for 3G networks, too. WiMAX initially may be deployed as a wireless backhaul solution, but will be upgraded to a mobility application starting 2007–2008, once the 802.16e standard is ratified and WiMAX-capable client devices enter the market, marking a ramp-up in the market.

WiMAX can complement existing and emerging 3G mobile and wireline networks, and can play a significant role in helping service providers deliver converged service offerings that can be accessed using a broad range of devices on a wide variety of networks.

At the technical level, 3G and WiMAX solutions fit well together, by providing different capabilities while allowing for seamless integration. 3G technologies have evolved over many years to become highly spectrally efficient, allowing operators to take advantage of costly spectrum dedicated to mobile services. 3G CDMA technologies such as W-CDMA (UMTS) and CDMA2000® 1xEV-DO provide high throughputs in low bandwidths — 5 MHz and 1.25 MHz, respectively.

Radio Technology: Push in the Right Direction

Radio technology and standards are still very much in a period of active and successful development. Their performance is still limited by the amount of cost-effective computing power that can be deployed. Before the advent of digital processing, radios were designed entirely with analog circuitry. As advances in the cost and scale of CMOS technology provided digital processing power, digital signal processing (DSP) began to play a major role in overall communication system designs. Ever-improving DSP techniques have enabled improvements in communications consistent with the predictions of Moore's Law.

The problem of reconstructing a data stream from a received encoded radio wave is solved with complex signal processing mathematics. As processors get faster, more complex algorithms become cost-effective, and the performance of digital radio systems improves. Although there are certainly theoretical limits to such improvements, there is still quite a way to go to reach them, and therefore we should expect continued technology evolution.

Today, we are experiencing the power of DSP techniques through many wireless radio frequency (RF) communication applications. Wireless wide area networks (WWAN, or cell phones), wireless local area networks (WLANs), and wireless personal area networks (WPANs) all employ sophisticated communication techniques.

Some of these techniques include complex modulation schemes, powerful new error-correcting codes, and decoding algorithms to combat the effects of channel fading, and so on. All these techniques are being enabled, cost-effectively, by the increasing capabilities of digital processing, as radio technologies continue to fulfill Moore's law.

Radios that use the spectrum in different ways are also being considered. One example is a technique called Ultrawideband or UWB. In this approach, vast bands of spectrum are used at extremely low power to transmit information very fast but only over very short distances. This approach allows multiple uses of the same spectrum because the UWB signal is so weak that existing users of the spectrum see it as a small amount of noise that can be ignored.

Closer to the application layers, there is also much ongoing work aimed at providing better user experiences. For example, it will become possible to build phones that can seamlessly roam from 802.11 systems to 802.16 systems to cellular systems. This could permit a handset to operate as an extension, and also behave as a cellular phone when its user leaves the building.

About the only thing that is certain about the directions radio technology will take over the coming years is that it will become increasingly digital, increasingly "smart," and continue to change. The biggest impediment to the deployment of these new technologies may well be the inadvertent barriers created by old regulations that preclude deployment. The technology challenges will therefore be shared among the technologists, the regulators, and those who wish to deploy useful technologies.

WiMAX Poised for Take-Off (or Maybe Not)

WiMAX, a technology that can provide wireless broadband coverage over a distance of 8 to 10 mi versus Wi-Fi's 300 ft, is about to take off. The industry approved its first WiMAX standard in July 2005. And when more flexible versions of WiMAX are approved in 2006 that will allow for WBA anywhere, sales of WiMAX gear are expected to shoot through the roof.

However, in some ways, WiMAX is not really a new technology; rather, it is an evolution of BWA, which has been used throughout Africa and the Middle East for over ten years now, providing voice and data services in both rural and urban areas where there was no telecom infrastructure, or where the infrastructure was old or saturated and the replacement or upgrade would have been prohibitively expensive.

However, because WiMAX is just the next generation in the evolution of BWA, it doesn't mean that the new opportunities for incumbent and emerging operators are not new. Also, it doesn't mean that the opportunities WiMAX presents to new and existing operators are not worth exploring. Some of the strong indicators that the BWA industry is destined for success in the times to come are the following:

- Scalability — Roaming from any access network to any other access network (2G, 3G, 4G, Wi-Fi, WiMAX, Bluetooth, satellite, Ethernet).
- Standard handover interfaces — Interoperability between different vendor equipment.
- Cross-layer solutions — Extensions to layer 1 and layer 2 functionalities to optimize higher-layer mobility architectures (MIPv4, MIPv6, SIP).
- QOS guarantees during handover — No disruption to user traffic, extremely low latency, signaling messages overhead and processing time, resource and route setup delay, near-zero handover failures, and packet loss rate.
- Security — User maintains the same level of security when roaming across different access networks.

Is WiMAX Secure?

WiMAX offers a level of security on par with any other high-security wireless network. WiMAX is more secure than Wi-Fi, and similar to the security levels of cellular networks.

Also, with WiMAX capable of acting as a backhaul for the network, its architecture makes it one of the few robust secured wireless networks. In this way the entire value chain between client and server is secured, so much so that enterprises do not really need to worry about security at all.

Positive Spectrum Environment

In other schemes, smart or agile radios are being considered that can listen to spectrum bands to see if they are currently in use by their official owner, and if not, opportunistically use them to transmit data. Such radios could allow much more total efficiency in the use of spectrum than our more typical fixed allocations permit. Many of these new radio approaches, however, will require that regulators redesign their spectrum regulations to be more flexible. It will be critical for regulators to understand how they can accommodate such radically new approaches lest these radios be precluded from operating in their countries.

As in the previous appearance of BWA, spectrum issues were spoilsports though much has changed since then. NextLink spent $695 million purchasing LMDS licenses from WNP in 1999, and LMDS license costs are estimated to have been $40 per head of population at that time. By contrast, for WiMAX, the unlicensed spectrum is free, and 3.5 GHz is going for under $5 a head, and sometimes under $1.

In spectrum terms, WiMAX operators have the option of unlicensed 5 GHz bands, though these bring some QoS risks and make it impossible to exploit the mobile and NLOS potential of 802.16 in the lower frequencies. A far more attractive choice is 3.5 GHz (or 2.5 GHz MMDS in North America), though the potential opening up of further, lower bands for broadband wireless could create still-greater opportunities in the coming years. For U.S. operators, the shortage of MMDS is a problem, although its once-restricted usage is almost certain to be liberalized. All three, Sprint-Nextel, BellSouth, and Clearwire, who have current licenses for MMDS, however, have the financial and marketing resources to build a national network and hardly notice the $3 billion bill, because the spectrum is already paid for (and was a low-cost purchase to begin with, unlike the LMDS licenses that Teligent and the others acquired).

In Europe, many countries are auctioning 3.5 GHz licenses this year or did so in 2004, and as regulators look to extend broadband access to most of the population in line with EC and national guidelines, the trend is to offer the licenses at relatively low cost. For instance, the national licenses for Austria cost about €700,000, compared to the €5 million many bidders had originally factored into their business plans. In other countries, such as Ireland and some of the new EU entrants, regional licenses can be had for five-figure sums. This has opened up the market to new operators, such as Altitude in France and WiMAX

Telecom in Austria, which do not need to raise vast sums to become national telecommunications players.

Forecasts and Global Trends

Telecommunication

Telecommunication has started looking up subsequent to a long consolidation phase after the telecom bubble bust. This time the thrust is on investments that can provide returns in form of increased efficiency of the telecom setup or increase in the addressable market. The trend of spending on software-based systems validates to the first aspect as software-based systems substantially increase efficiency. To increase the addressable market, two approaches are used, one to reach more new prospects by, for example, targeting suburban and rural customers not being offered services so far, or to increase the number of prospective services offered to existing clients.

The international telecommunications market is expected to continue its double-digit growth from 2004 to reach over $2 trillion by 2008, according to the Telecommunications Industry Association (TIA) 2005 Telecommunications Market Review and Forecast.

Overall spending on telecommunication in the five regions covered in the report — Canada, Western Europe, Eastern Europe, Latin America, and Asia Pacific — will grow at an estimated 10.6 percent compound annual growth rate (CAGR). The principal drivers for this growth, said TIA, are improving economic conditions throughout the world, a growth in infrastructure equipment investment, and demand for mobile devices and wireless services.

Telecom spending in the next few years will revolve around software-based intelligent systems for increasing efficiency, outsourced services for efficient network planning and optimization, and last but not the least, the access technology.

The telecommunications market is also due for a shift in leadership positions, the increasing clout of Asian players being increasingly evident.

On the revenue side we would likely see the gap between wireless and fixed wireline service providers narrowing further; also, it is expected that by 2007 wireless revenue may cross the wired or fixed revenue. Data services offered by wireline operators are the key reasons that have delayed the inevitable so far; the number of wireless subscribers crossed that of fixed subscribers in 2002.

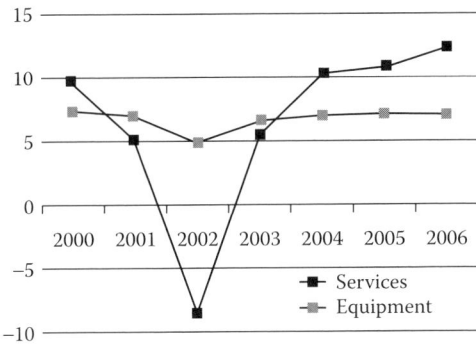

Figure 11.3   Global telecommunications services and equipment revenue growth.

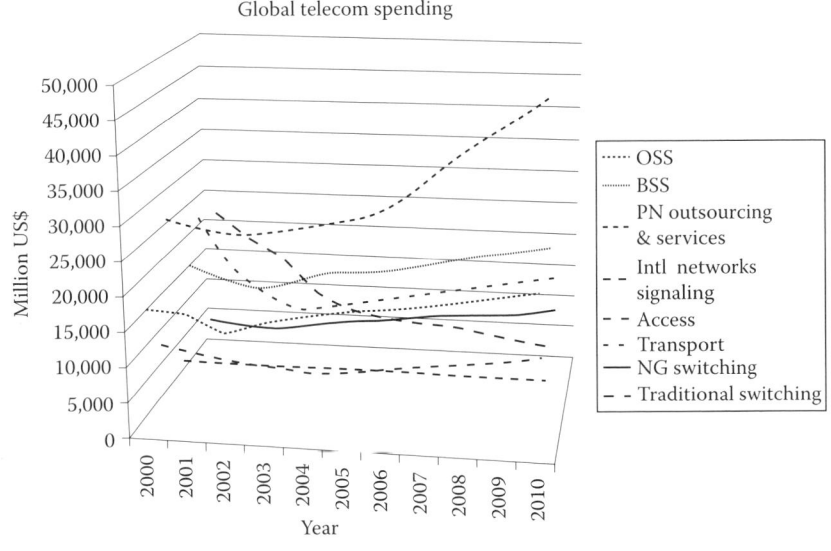

Figure 11.4   Global telecommunications infrastructure spending trends.

The future of wireline or fixed services depends substantially on data and broadband Internet access services.

Mobile

There has been phenomenal growth in the number of Internet and mobile users. Both technologies have grown at lightning speed, and they seem to be heading in one direction only — up.

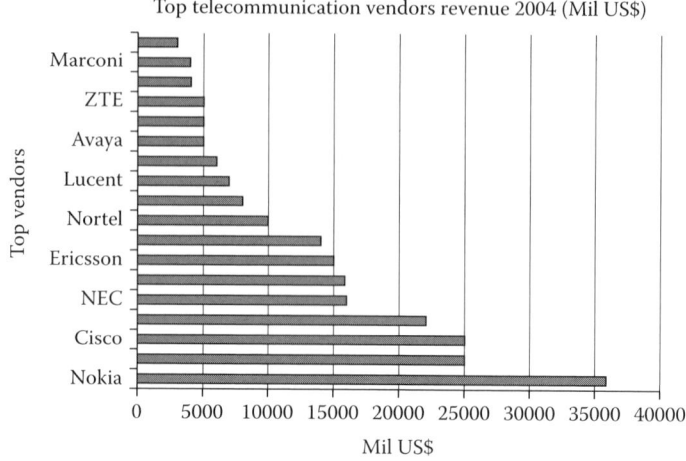

Figure 11.5   Global telecommunications vendor revenue trends.

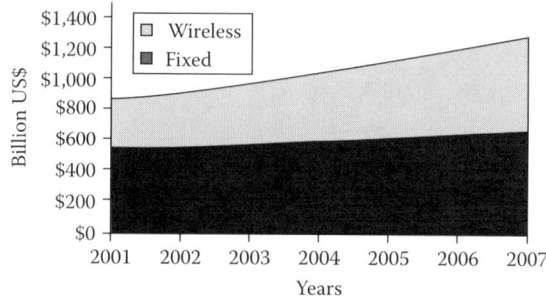

Figure 11.6   Global fixed and wireless telecommunications services revenue trends.

The number of wireless subscribers is growing faster than the number of landlines in each region and is expected to reach 1.9 billion in 2008, outnumbering landline subscribers by 69.1 percent. As wireless penetration grows, the average penetration rate for all regions is expected to reach 44 percent by 2008.

As more and more people embrace broadband and wireless data, the revenue generated by data services will rise substantially (by nearly 30 percent) in comparison to voice (less than 2 percent). Further, with wireless VoIP gaining a foothold later, the trend would be far more antagonistic to voice, which may even show negative growth.

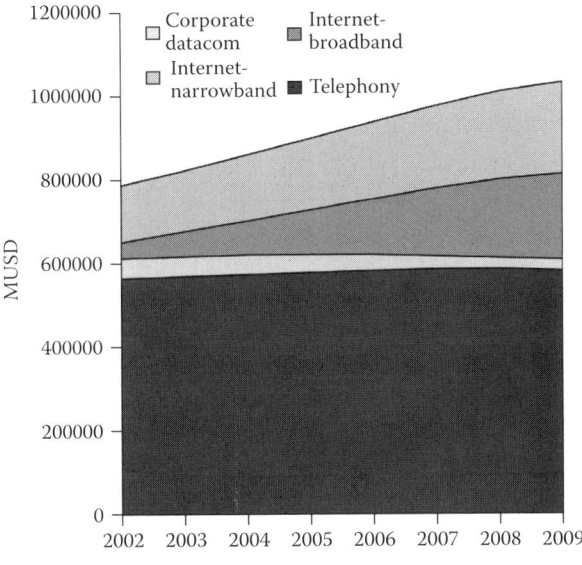

Figure 11.7 Global wireline telecommunications revenue trends.

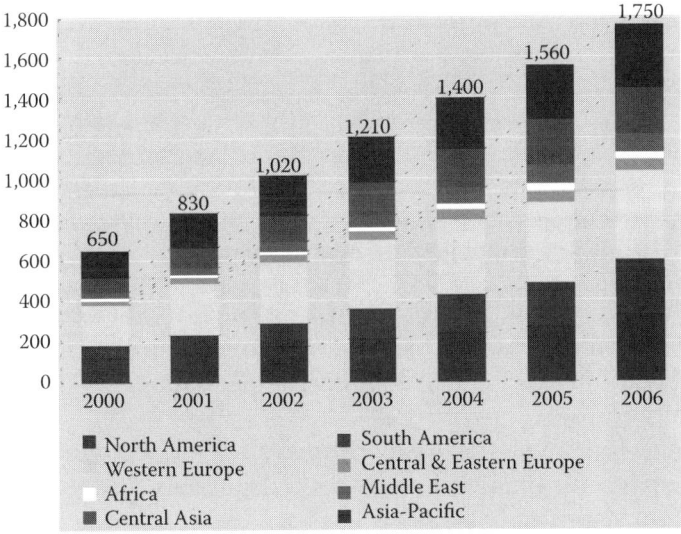

Figure 11.8 Global mobile users by region (in millions).

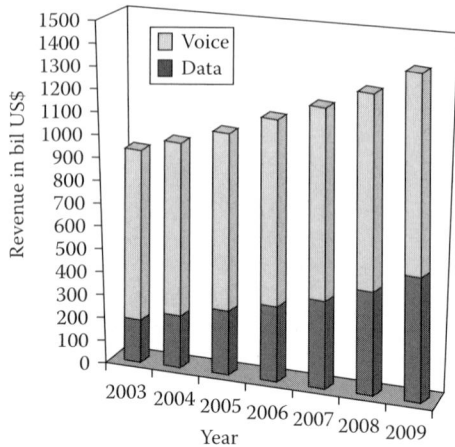

Figure 11.9   Global data and voice revenue trends.

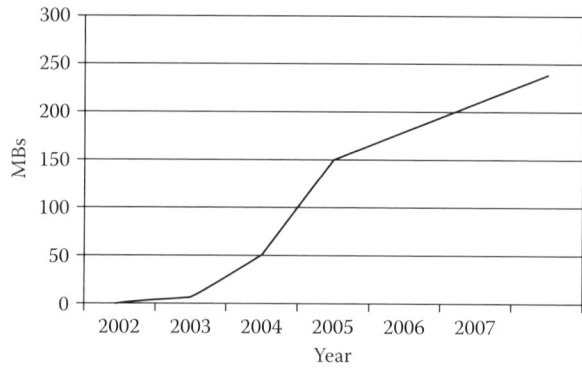

Figure 11.10   Per-user mobile data demand trends.

Another interesting aspect about this voice to data competition is that now mobile telephony has also started drifting toward data-based growth. With mobile data requirement escalating, so will wireless communication.

Demand for mobile data is good news for mobile handset vendors as the demand for high-end high-speed data-enabled phones will also increase substantially.

The balance of spending for telephony has now moved to wireless, and wireless telephony has the largest share of the consumer's wallet — and the only one growing. Will it be broadband's turn to go wireless next?

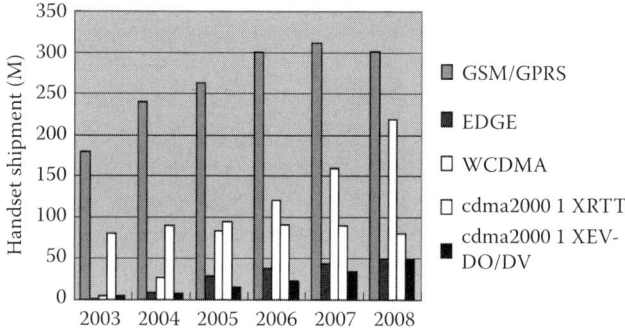

Figure 11.11 High-speed data enabled mobile phones shipment trends.

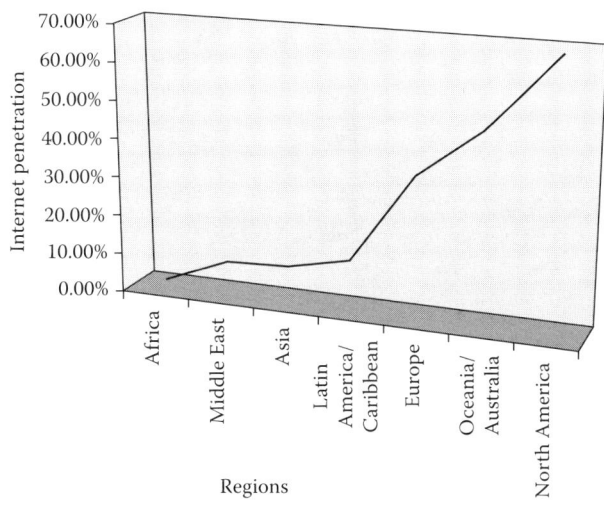

Figure 11.12 Internet penetration — by region, 2005.

Internet

Internet growth has continued, and the number of Internet users is nearly one billion.

Asia has the highest number of Internet users, and is followed closely by Europe. In Africa and Latin America, Internet usage has not reached substantial levels.

Most affected by the digital divide is Africa, with less then 2 percent share of the worldwide Internet users although representing 19 percent of the world's population.

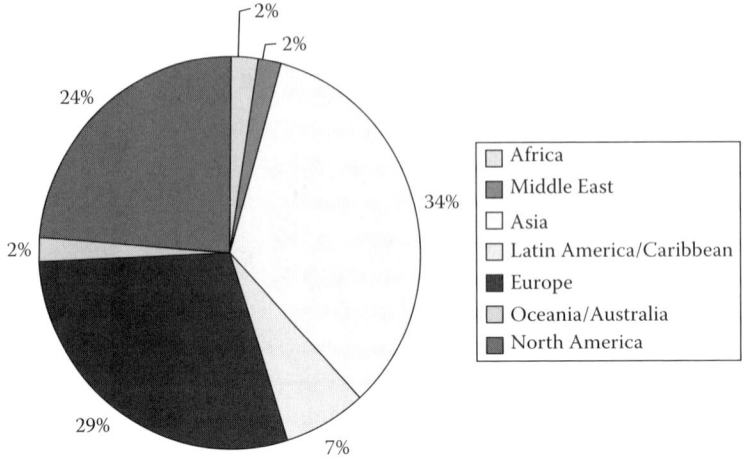

Figure 11.13   Internet users — by region, 2005.

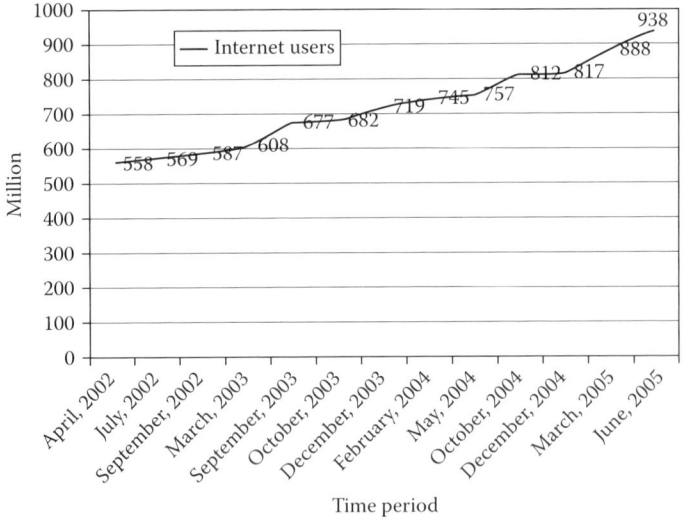

Figure 11.14   Internet user trends.

Broadband

High-speed broadband access will be a principal driver of equipment revenue in the next four years, helped by increased government

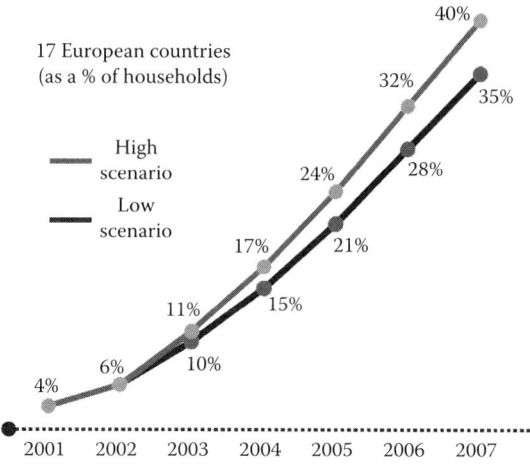

Figure 11.15   Broadband penetration — Europe.

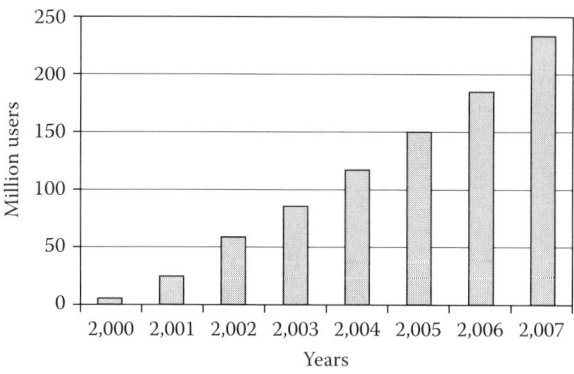

Figure 11.16   Broadband Internet user trends.

support and a stronger economic environment, according to TIA. The group expects broadband access revenue to triple between 2004 and 2008, from $33 billion to $101 billion.

As the broadband market expands, the need for infrastructure to support the traffic will revitalize the network infrastructure equipment market. TIA expects equipment spending to increase at an 8.1 percent CAGR, rising from $238 billion in 2004 to $325 billion in 2008.

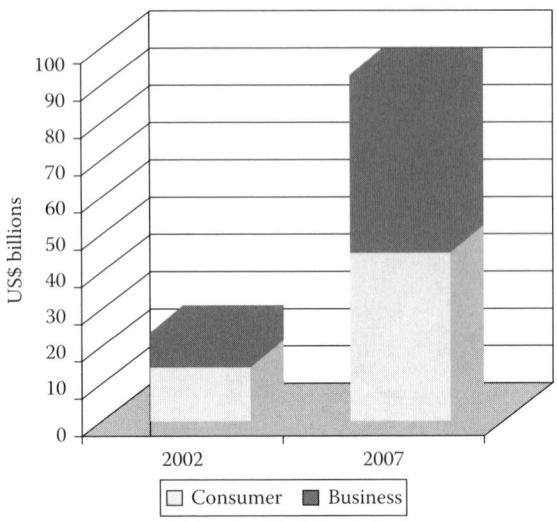

Figure 11.17   Broadband Internet revenue trends.

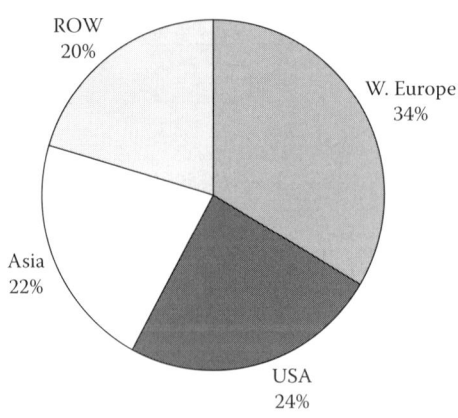

Figure 11.18   Broadband Internet — by region (2005).

BWA Industry

We expect BWA connections to expand globally at a very healthy 27 percent CAGR between now and 2010. This is significantly greater than the prevailing single-digit growth rate of the telecoms industry as a whole, and it does not take into account any effect WiMAX may have on expanding the market.

Broadband - according to technology - 2007

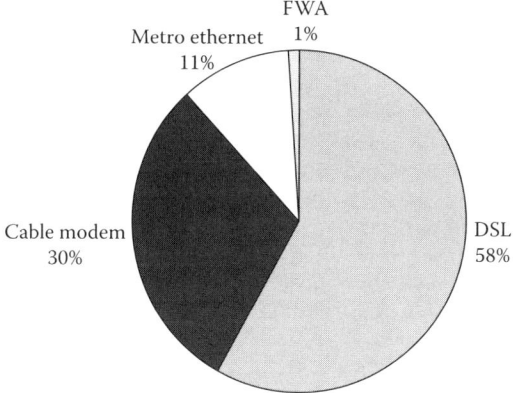

Broadband - according to technology - 2002

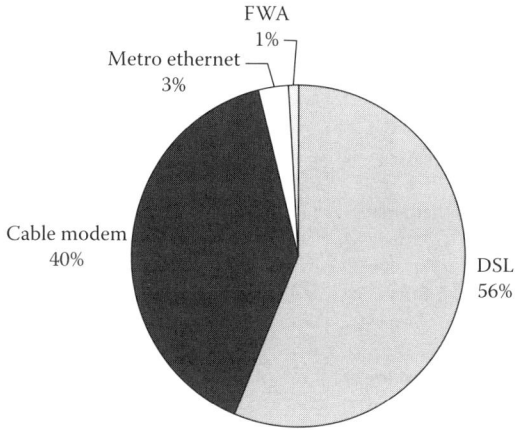

Figure 11.19   Broadband Internet trends.

WiMAX

WiMAX will succeed globally, albeit unevenly. WiMAX will succeed in every geographic market, but for different reasons. In emerging markets, operators are interested in using WiMAX for low-cost voice transport and delivery. In developed markets, WiMAX is all about broadband Internet access. Overall, the markets without any fixed infrastructure offer the greatest opportunities. WiMAX will become a disruptively inexpensive means of delivering high-speed data.

As the distinctions between fixed and mobile services blur, a chaotic mix of large fixed and wireless providers will pursue WiMAX deployments. The local and regional wireless ISPs are likely to be acquired

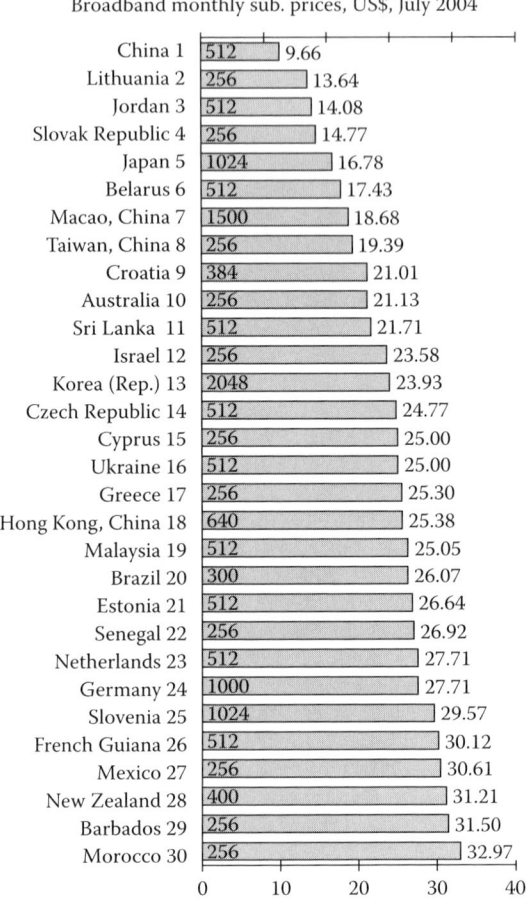

Broadband monthly sub. prices, US$, July 2004

Country	Speed	Price
China 1	512	9.66
Lithuania 2	256	13.64
Jordan 3	512	14.08
Slovak Republic 4	256	14.77
Japan 5	1024	16.78
Belarus 6	512	17.43
Macao, China 7	1500	18.68
Taiwan, China 8	256	19.39
Croatia 9	384	21.01
Australia 10	256	21.13
Sri Lanka 11	512	21.71
Israel 12	256	23.58
Korea (Rep.) 13	2048	23.93
Czech Republic 14	512	24.77
Cyprus 15	256	25.00
Ukraine 16	512	25.00
Greece 17	256	25.30
Hong Kong, China 18	640	25.38
Malaysia 19	512	25.05
Brazil 20	300	26.07
Estonia 21	512	26.64
Senegal 22	256	26.92
Netherlands 23	512	27.71
Germany 24	1000	27.71
Slovenia 25	1024	29.57
French Guiana 26	512	30.12
Mexico 27	256	30.61
New Zealand 28	400	31.21
Barbados 29	256	31.50
Morocco 30	256	32.97

Figure 11.20 Broadband Internet cost — by country.

as large carriers, particularly fixed, turn their attention to rural areas and enterprise accounts. For example, in the United States we have seen ISPs compete successfully in moving enterprises from T1 lines to wireless lines; fixed carriers will eventually be forced to respond, either through similar deployments or by acquisitions.

WiMAX will evolve in two stages. The first stage will begin in 2006 with products that cost and function similar to current BWA equipment. The total fixed wireless market will not expand as a result of WiMAX; what we will see is a gradual migration of purchasing behavior from proprietary equipment to WiMAX equipment. Operators will be leery

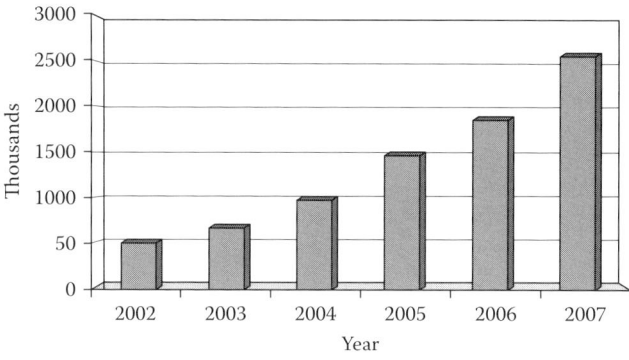

Figure 11.21   Global fixed wireless broadband trends.

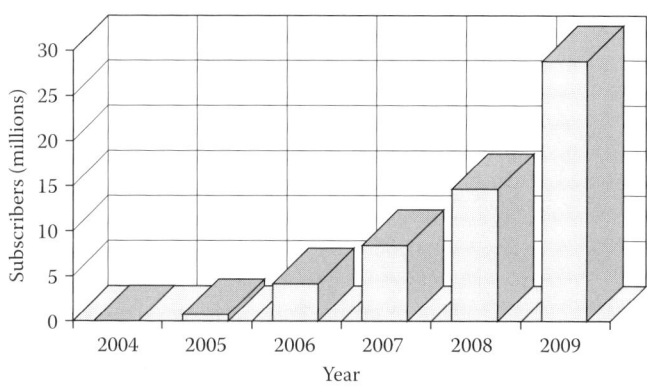

Figure 11.22   Global wireless broadband trends.

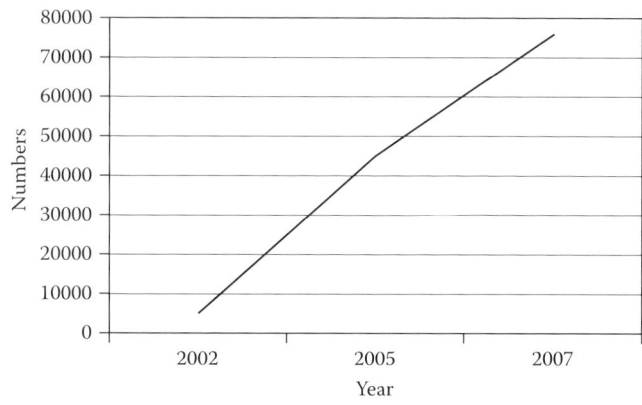

Figure 11.23   Global Wi-Fi hot spot trends.

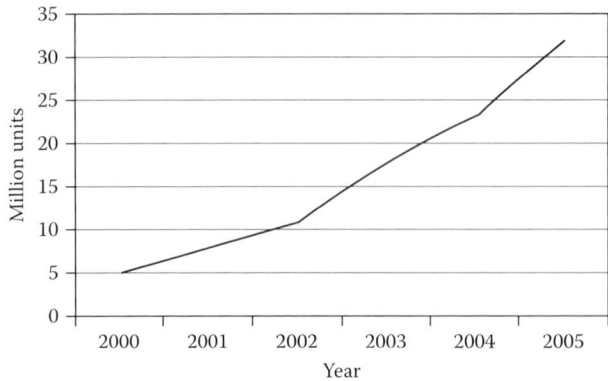

Figure 11.24   Global Wi-Fi CPE shipment trends.

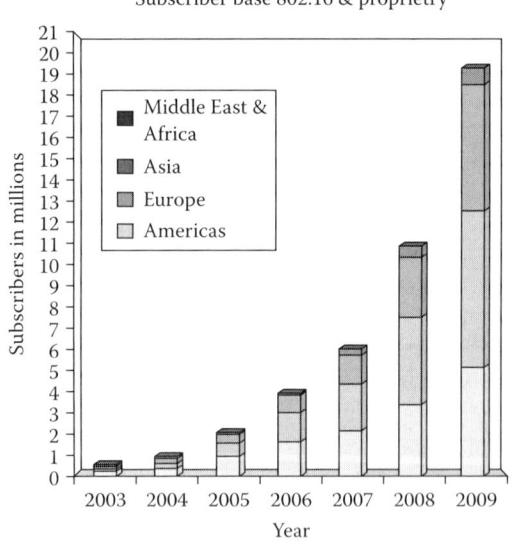

Figure 11.25   Global WiMAX broadband lines trends.

of adopting WiMAX equipment until prices drop to the point where they cannot afford to ignore WiMAX, which should occur in late 2005.

At about this time, we will see the beginning of the second stage of WiMAX: the birth of metro-area portability. Once 802.16e is approved, laptops and other mobile devices may be embedded with WiMAX chipsets, so that the users can have Internet access anywhere within WiMAX zones. If this sounds a lot like 3G, in many ways it is.

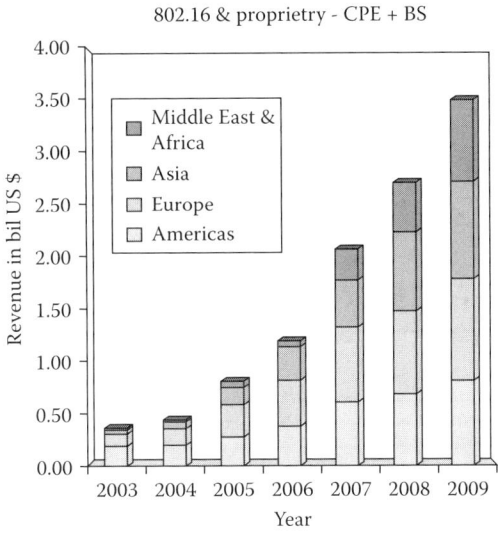

Figure 11.26   Global WiMAX broadband equipment revenue trends.

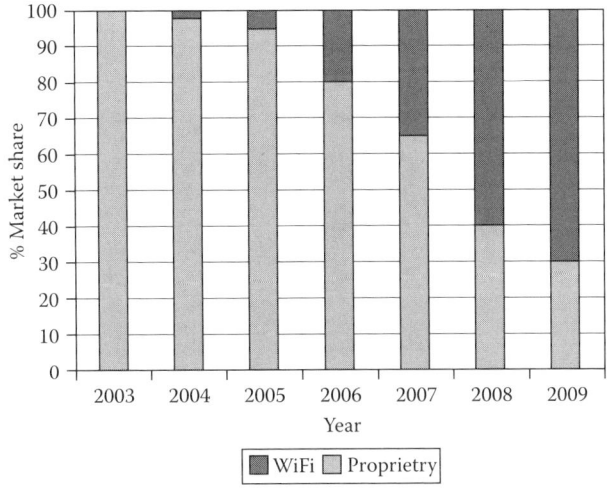

Figure 11.27   Global market share — WiMAX and proprietary.

The second stage of WiMAX could be very disruptive to 3G operators and could drive a round of WiMAX network overlays in urban areas. Nevertheless, this will not happen until 2006 at the earliest. As shown in the following figures, WiMAX (stages one and two) and Wi-Fi will complement one another.

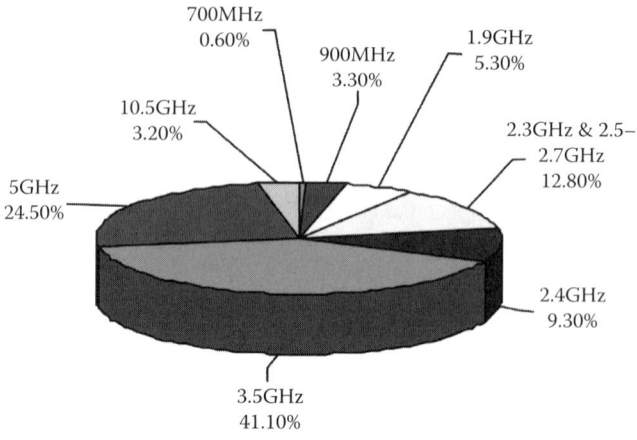

Figure 11.28 Global WiMAX market share — by frequency.

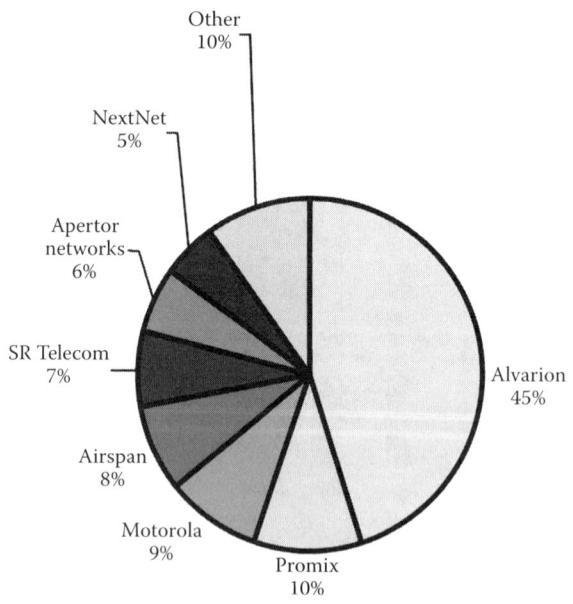

Figure 11.29 Global WiMAX market share — by vendor.

Chapter 12

Analyzing the Model

The customer will determine the development and success of the WiMAX business opportunity, so to capitalize on WiMAX the players must let the customer and market drive technology development — technology cannot dictate customer demand. WiMAX might have a major impact on the value chain and end users, but the mechanism is unclear and depends on the parameters given in the following text as well as on the ability of WiMAX to deliver as promised.

Because of high market and technological uncertainty regarding the various stages of WiMAX development, the different business models give high and varying discount rates for WiMAX business cases. WiMAX may be hyped, and there is still substantial uncertainty and risk related to the following:

- Technology, standards, and performance
- Value to customers
- Business models and business cases
- Alternative solutions and technologies
- Market dynamics and competitor behavior/response

Users will not pay for a new generation of technical complexity, no matter how much money the industry puts into its development, but they will pay for better products and services that bring them value.

WiMAX: Scores Higher

WiMAX is one of ten or more broadband wireless access (BWA) technologies, of which all have pros and cons. What makes WiMAX appealing is that unlike other BWA technologies, there are many compelling cost-, application-, and scope-related advantages of using WiMAX products for access:

Wider access scope: WiMAX adopts orthogonal frequency division multiplexing non-line-of-sight propagation technology to provide broadband access for residents or enterprises for a surrounding area of more than 10 mi (theoretically, it can be up to 31 mi). In areas where wired resources are scarce and of poor quality, the advantage of WiMAX access is particularly apparent.

Competitive costs: WiMAX products are based on the 802.16 standards and have to pass the consistency certification conducted by the WiMAX Forum to ensure the interconnection and interoperability of equipment of different manufacturers. Participation of such chipmakers as Intel will greatly reduce the costs of WiMAX products. In addition, because WiMAX is a wireless access technology, operators do not need to invest in cable installation, the construction period is short, and capacity expansion and removal is flexible and convenient. All these factors allow operators to cut capital investment, speed up capital turnover and recovery, protect investments already made, and cut business risks.

Higher bandwidth: WiMAX products provide higher width than conventional access modes, which make them more suitable for application in high-traffic hot spots such as enterprises, hotels, Internet bars, and IP supermarkets. WiMAX products are more suitable for providing services, such as multimedia, VoD, and videoconferencing, which require higher bandwidth. In this way, operators can effectively solve the last mile access bottleneck.

Secure transmissions: The WiMAX system has QoS and encrypted transmission functions, which ensures QoS and secure transmission of information in the last mile.

SWOT Analysis

Strengths

Based on proven OFDM techniques (inherent robustness against multipath fading and narrowband interference):

Low cost to deploy and operate (~$33 per home versus $300–600 for DSL)

High speed (75 Mbps) and long range (50 km)

Adaptable and self-configurable

Centralized control in MAC enables simultaneous, varied QoS flows

Weaknesses

Currently high power consuming (still far from penetrating portable mobile devices)

Mobility not yet fully specified — could become complex to implement

Opportunities

High-speed wireless infrastructure

Cellular infrastructure for converged networks

Last mile solution for broadband wireless access

CAPEX of ~$3.7 billion needed to penetrate

97.2 percent of U.S. homes (In-Stat/MDR)

Threats

DSL/ADSL technologies widely deployed

Cellular penetration is very high, and growing

Possible wide deployment of 3G

Widespread success of 802.20 standards

Where WiMAX Fits In

WiMAX will be in principle used by incumbent local exchange carriers (ILEC), competitive local exchange carriers (CLEC), Wireless ISPs (WISPs), mobile operators, and enterprise/public bodies. All of these users have a lot of diversity; hence, along with features of various flavors of WiMAX, it is expected that we will see a deluge of models.

Unlike a simple business model, every WiMAX model would be an aggregate of a number of submodels. Basically, there will be different submodels based on the following:

1. Mobility — Fixed, nomadic, and mobile
2. Equipment CPE — Outdoor and indoor (also referred as self-installed)
3. Ownership — Public, private, and community
4. Application — Backhaul, last mile, and end to end
5. Regulations — Licensed and license-exempt
6. Rollout — Add-on and from the ground up

Mobility: Fixed, Nomadic, and Mobile

The real disruptive potential of WiMAX lies in mobility. Without mobility and incorporation into laptops and handsets, 802.16X remains just a more cost-effective step forward from traditional fixed wireless solutions, based on standards. Mobility raises the possibility of creating networks with full broadband speed over regional or national areas, fulfilling many of the half-kept promises of public Wi-Fi and cellular 3G.

The mobile version of the standard, 802.16e, is not even ratified yet, and it will be a year before we see even first-stage products starting to appear.

At its simplest level, WiMAX is intended to provide definitive IP standards for a carrier-class solution that can scale to support thousands of users with a single base station, and provide differentiated service levels. By enabling IP standards-based products with fewer variants and larger volume production, WiMAX should drive down the cost of network equipment and make broadband wireless a viable alternative to wireline technologies. Soon, a single base station sector will provide enough data rate to simultaneously support more than 60 businesses with T1-type connectivity and hundreds of homes with DSL-type connectivity.

As a result, CLECs will be able to provide a real broadband alternative using their own infrastructure, ILECs will be able to deploy high-speed Internet access in regions where wired connections are not profitable, and WISPs presently using Wi-Fi or any other technologies would be able to extend their existing services.

Equipment CPE: Outdoor and Indoor

The technology is expected to be adopted by new, as well as different, incumbent operator types, for example, wireless Internet service providers (WISPs), cellular operators (CDMA and WCDMA), and wireline broadband providers. As the self-install or indoor CPE does not need

Table 12.1 Internet Access Customer Acquisition Cost

Outdoor Antenna Installation	Indoor Antenna Installation
$300 to $500 truck role	$0
$400 CPE (subsidized)	Self-installed $250 CPE (subsidized)
Totals = $800 @ $20/month equals 40 month break even	Totals = $250 @ $20/month equals 12 month break even

Note: ARPU = $20/month.

truck rolls, it definitely will be more cost-effective. But for applications such as backhaul or in which the high throughput at long distances is key, as in the case of rural or suburban areas, CPE with outdoor antenna can be deployed. Again, depending on the application, models will differ.

Self-install indoor CPE delivers a viable business case for residential BWA. Let us take two different cases with the customer having the same average revenue per user (ARPU) but one using indoor self-installed CPE and other using outdoor CPE (Table 12.1).

Indoor self-installed antennas imply faster breakeven.

Ownership: Public, Private, and Community

Entities such as weather carriers, companies, communities, or cities that are considering building and operating wireless WiMAX networks need to consider the array of financial and business planning models.

Among various WiMAX network ownership models, a few relevant to a metropolitan access network are a city-owned municipal WiMAX network to replace or enhance the existing telecommunications infrastructure; community-owned public/private WiMAX network for municipal, commercial, and residential use; and a private or cooperatively owned and operated WiMAX network that the city supports by buying service, through promotion or some other means.

Application: Backhaul, Last Mile, and End to End

Based on the application, WiMAX can be deployed in various configurations such as backhaul, last mile, or large-area coverage access. Backhaul uses point-to-point antennas to connect aggregate subscriber sites to each other and to base stations across long distances. Revenue generation is based on the data pipe as access is provided by other technologies. For last mile or large-area coverage access, revenue is based on penetration as WiMAX provides services to end users.

Regulations: Licensed and License-Exempt

WiMAX could form part of a wider wireless broadband strategy, comprising both licensed and unlicensed technologies, including GSM or CDMA, UMTS, proprietary broadband fixed wireless (LMDS, MMDS), Wi-Fi, and wireless mesh networks.

Licensed and license-exempt WiMAX solutions face common challenges related to government regulations, infrastructure placement, and interference. However, license-exempt solutions have more to prove in environments in which licensed solutions are seen as more stable and reliable.

A key difference between these two models is that investment may be higher in the licensed model as license fees are paid to the regulator for spectrum. Depending on various factors such as interference and infrastructure location constraints, sometimes the license-exempt model may not be a very good idea.

Rollout: Add-On and Ground Up

Depending on the present situation and strategy of the carrier, rollout of WiMAX services is done. For existing players it is generally add-on, in which new infrastructure is added on the existing one to enhance the services. In this case, first inventory analysis is done to find out the most optimum deployment. For new players it is ground up as the infrastructure is planned from the scratch.

Business Case

This section will provide a detailed business case analysis for WiMAX technology in fixed wireless applications in the sub-11-GHz frequency range. In this business case, operators offer services to both residential and small-to-medium business customers. Business case assumptions used are the typical parameters an operator would experience in most developed countries, where wireless access provides a competitive alternative to DSL, cable, and aggressively priced leased lines.

Considerations and Assumptions

An accurate business case analysis must take into account a wide variety of variables.

Demographics

Demographics play a key role in determining the business viability of any telecommunications network. Traditionally, demographic regions are divided into urban, suburban, and rural areas. In our analysis, a fourth area has been added called *exurban*. Exurban areas are primarily residential, and when compared to suburban areas are further from the urban center. They are located at the periphery of the urban municipality limit, with lower household densities.

Services

Following is a description of the services used in the business case with the assumed first year ARPUs. These ARPUs are competitive with or below current cable, DSL, and leased line services in most developed countries. The ARPUs are assumed to drop 5 percent per year after the first year.

In addition to high-speed Internet access, it is assumed the operator will also offer voice services to residential and SME customers. Other revenue sources include one-time activation fees and equipment rental fees for operator-supplied CPE. These fees are assumed to stay constant over the business case period.

Regulator-imposed taxes and tariffs are not included in the analysis, because these costs are generally passed on to the end customer and will therefore have little or no impact on the business case (Table 12.2).

Frequency Band Alternatives

In our analysis we will use the 3.5 GHz band for metropolitan area deployment and the 5.8 GHz unlicensed band for rural area deployment (Table 12.3).

Scenario for Business Case Analysis

For the business case analysis, three different scenarios are analyzed, which are summarized as follows (Table 12.4).

Table 12.2 Business Case Assumptions

End Customer	Service Description	ARPU First Year ($)	Other Revenue
Residential			
Residential Internet	A "best-effort" service (assume 384 kbps with 20:1 oversubscription)	30	$10/month for equipment lease and one-time $50 service activation fee
Residential POTS	VoIP Service	20 Optional	
Small Medium Business			
Basic service	0.5 Mbps CIR, 1 Mbps PIR	350	$35/month equipment lease fee and one-time $500 service activation
Premium service	1.0 Mbps CIR, 5 Mbps PIR	450	
Local access (POTS)	T1/E1	200 Optional	
Wi-Fi hot spot backhaul	1.5 Mbps CIR, 10+ Mbps PIR	650	$25/month equipment lease fee and one-time $500 activation fee

Capital Expense (CAPEX) Items: Base Station, Edge, and Core Network

The business case assumes a greenfield deployment and, as such, must allow for core and edge network equipment in addition to WiMAX-specific equipment. In this business case analysis, base station capacity is determined by using a 20:1 overbooking factor for "best-effort" residential services assuming 384 kbps average data rate and 1:1 for SME committed information rate (CIR) services. For the residential case this conservative overbooking factor should enable WiMAX subscribers

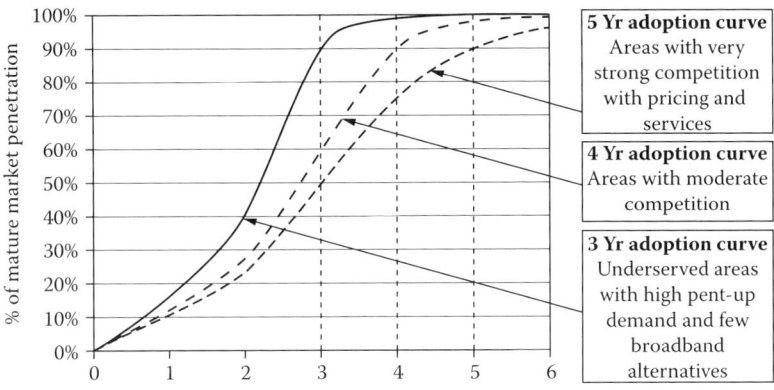

Figure 12.1   Market adoption curve.

to experience performance during peak periods superior to what many DSL and cable customers experience today.

The business case also assumes the deployment of a high-capacity point-to-point wireless backhaul connection for each base station to a point of presence or fiber node for connection to the core network. This can also be accomplished by means of leased T1/E1 lines in which case, rather than a capital expense, there would be an operating expense.

It is assumed that a spectrum license is obtained through an auction process at a cost of $0.01 per MHz point of presence (PoP) (Table 12.5).

CPE Equipment

WiMAX equipment manufacturers will be providing CPE hardware in a variety of port configurations and features to address the needs of different market segments. Residential CPEs are expected to be available in a fully integrated indoor self-installable unit as well as an indoor/outdoor configuration with a high-gain antenna for use on customer sites with lower signal strength. In the business case analysis, a percentage breakdown of each is assumed in accordance with the frequency band, cell radius, and propagation conditions that are likely to be encountered in the different geographic areas.

For both the residential and SME market segment, it is assumed that a percentage of customers will opt to purchase their own equipment rather than pay an equipment lease fee to the operator. This has the effect of reducing the CPE CAPEX and CPE maintenance expense.

Table 12.3 Frequency Alternatives

Band	5.8 GHz Band	3.5 GHz Band
Licensing	License-exempt spectrum	Licensed spectrum in Europe, Latin America, and Asia
Cost	n/a	Cost will vary from country to country depending on regulations, may also lease from existing license holder. Cost, therefore, could be CAPEX or OPEX
Spectrum	Up to 125 MHz available in the United States (may vary in other countries)	Assignments vary country by country
Allowable transmit power	The United States: Maximum power to antenna 1 W, Max EIRP +53 dBm (200 W); limits may vary in other countries	Per ETSI: 3 W (+35 dBm) maximum to antenna (may vary in countries outside Europe)
Interference control	Restrict deployment to less than half the available spectrum, use auto channel select and coordination between operators	Protected by license assignment, no two operators assigned the same frequency in the same geographic area
Base stations required	Higher base station capacity results in fewer base station sites to achieve area coverage	More base station sites to meet capacity requirements because of limited spectrum assignment
Indoor and outdoor CPEs	Can support indoor CPEs at customer sites within ~800 m from the base station, outdoor CPEs must be deployed elsewhere. Result: higher average CPE cost and higher average installation costs	Will support a high percentage of indoor CPEs in capacity limited deployments. Result: lower average CPE cost and lower average installation cost

Note: n/a = Not applicable.

Table 12.4 Business Case Scenario

	Scenario
Geographic area description	Major city/metropolitan area
Market segment	Residential, SME, and Wi-Fi backhaul
Size	125 sq km
Population	~1,000,000
Residential density	6000 HH/sq km in urban center, 1500 HH/sq km in suburban area, 500 HH/sq km in exurban area
Total households	~390,000
Total SME	~24,000
Adoption rate	4 years
Frequency band	Licensed 3.5 GHz band
Channel BW	3.5 MHz, FDD
Assumed spectrum	28 MHz (2 × 14 MHz)

It also however, reduces operator revenues derived from equipment lease fees. Because of this interrelationship, the impact on the payback period is not significant.

The business case analysis assumes that the price of residential terminals will drop by about 15 percent per year because of growing volumes and manufacturing efficiencies and lower-volume business terminals will drop by about 5 percent per year.

The CPE costs used in the business case analysis are summarized in Table 12.6.

Table 12.5   CAPEX for Network Infrastructure

Description	Details	Comments
WiMAX equipment	$35K per base station for 4-sector/channel configuration	+$7K per channel for additional channels/sectors
Other base station equipment	$15K	Covers any necessary cabinets, network interface cards, etc.
Backhaul link	$25K for a PtP microwave link	Allows for at least one multiple hop in rural areas
Core and edge equipment	$400K	Router or ATM switch, NMS, etc.
Spectrum license	Assume $0.01 per MHz PoP	Assumes license is acquired via an auction process or other upfront investment
Base station acquisition and civil works	$50K average per base station	Includes indoor and outdoor site preparation, indoor to outdoor cabling, etc.

Table 12.6   CPE CAPEX

CPE Type	Year 1 CAPEX ($)	Annual Price Reduction (Percent)	Assumed Percentage of CPE Units Provided by Operator
Residential indoor self-installable CPE	250	15	80 for scenarios 1 and 3, 60 for scenario 2
Residential outdoor CPE	350	15	
Small business terminal	700	5	50
Medium business terminal	1400	5	
Wi-Fi hot spot terminal	300	5	20

Table 12.7   OPEX

OPEX Item	Business Case Assumptions	Comments
Sales and marketing expense (including customer technical support)	20 percent of gross revenue in year 1 dropping to 11 percent in year 5	Higher percentage or revenue in early years to reflect the fixed costs associated with these expenses, fifth-year levels are consistent with those for a mature stable business
Network operations	10 percent of gross revenue in year 1 dropping to 7 percent in year 5	
G & A	6 percent of gross revenue in year 1 dropping to 3 percent in year 5	
Equipment maintenance	5 percent of CAPEX for base station equipment 7 percent of operator-owned CPE CAPEX	Reflects higher maintenance costs associated with maintaining remotely located equipment
Base station installation and commissioning	About $3K for 4-sector base station	One-time expense
CPE installation and commissioning	Varies with market segment	Offset by one-time activation fee charged to customer
Base station site lease expense	$1500 per month per base station	Space for indoor equipment plus per-antenna lease fee for outdoor units
Customer site lease expense	$50 per month average for SME market segment	Not applicable for residential market segment
Allowance for bad debts and churn	12 percent for residential segment 3 percent for SME segment	—

Table 12.8   Market Summary

Spectrum		Deployment	
Frequency band	3.5 GHz	WiMAX base stations deployed	63
Channel bandwidth in MHz	3.50	Aggregate payload in Mbps	2,470
Spectrum required in MHz	28	Coverage area in sq km	125
Addressable market		Average data density in Mbps/sq km	20
Households covered	388,513	Population in coverage area	1,010,134
Businesses covered	24,312	Assumed mix — CPE	
Market penetration	4-year curve	Percentage of indoor residential CPEs	80 percent
Market adoption curve	4 years	Percentage of residential CPEs supplied by operator	60 percent
Residential market	6.3 percent	Percentage of SME CPEs supplied by operator	50 percent
Residential voice services	23 percent	ARPU price erosion	5 percent
SME market	7.8 percent	Average number of subscribers per base station	423
SME voice services	52 percent	Total CAPEX per subscriber	$456
Wi-Fi hot spots backhauled	30	Total CAPEX in millions of dollars	$12.2
Net present value (in millions of dollars)	**$12.8**	**Internal rate of return (IRR)**	**107 percent**

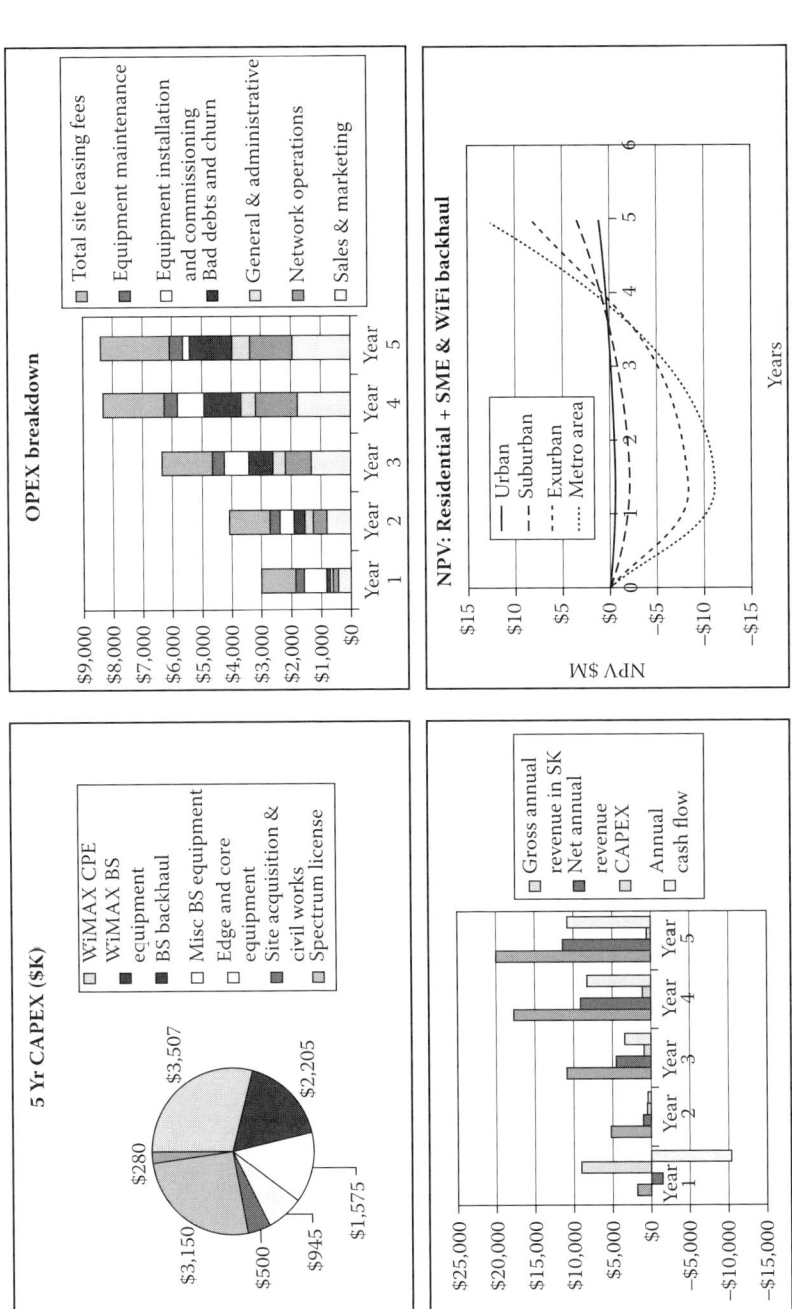

Figure 12.2 Financial summary.

Chapter 13

Planning the Strategy

In trying to come to grips with today's market and technology challenges, not only do telecommunications companies face massive difficulties in developing a viable strategy for their company but other players such as regulators, third-party service providers, equipment vendors, and end users also find themselves on shaky ground.

Even though different questions baffle different participants, a common theme more often than not can be discerned. In the early 1990s, doubts centered around the Internet, mid-1990s came with confusions related to mobile telephony, the late 1990s created much distress due to broadband, and now once again everyone wants to know what and what not to do to ride the broadband wireless access (BWA) bandwagon.

Big successes and bigger failures in the past have made the stakes much higher. Nobody wants to be left behind. Technologists and their business counterparts in corporate circles are trying to come up with a winning formula.

It may seem unlikely that a simple approach to such a complex problem exists. However, let us try to find a direct path to success for operators, users, regulators, and vendors associated with BWA, with an obvious focus on WiMAX, considered the biggest opportunity of our era.

Management Summary

In trying to get a handle on today's market and technology challenges, many telecommunications companies struggle to develop a viable business strategy for their company.

Top management in many telecommunications companies is busy finding answers to questions such as how to cope with increasing traffic volume? How to create sustainable stakeholder value? How to manage the ever-increasing variety of services that is demanded by divergent consumer markets? How to handle new technologies?

Many of these questions point toward a basic dilemma that any experienced professional must have faced many times in his or her career, namely, whether to grab the latest technology to corner the first-mover advantage or wait for others to make the first move and burn their fingers.

The successful telecommunications operator of the future will definitely embrace enabling technologies in a timely fashion, after having meticulously assessed where to go and how far to reach beyond the traditional value chain.

To be successful in identifying a new business opportunity and planning a detailed strategy to take advantage of the opportunity presented, a careful analysis of the value chain, evaluation of different services portfolios and, finally, selecting the most promising business model is necessary; so is the calculated risk taking.

So far, so good; but unfortunately, sometimes choices are not clear-cut. Often, organizations find more on the plate than they can chew.

In the present scenario, it is not enough to decide whether or not to go ahead with a particular technology, say, WiMAX. The question is far more complex, and hence the underlying strategy varies widely for different players, depending on its present status, the market, and many more parameters. The dynamic environment does not make the situation any easier. Strategy development in this new and rapidly changing environment is certainly a daunting task!

Strategy Development: An Art, Not a Science

The following text describes the fundamental logic for the creation of new, state-of-the-art strategies for successful performance that meets the market challenges described earlier. As this subject in itself is outside the scope of this book, we will give a broad overview before considering its relevance to WiMAX.

The main steps in this endeavor are the following:

- Strategy development: Conception of strategic options by assessing market potentials from a financial as well as a technical point of view. A suitable method is value chain analysis.
- Strategy evaluation: Testing these strategic conceptions and hypotheses for plausibility and feasibility, which can be a check against the market challenges and against the corporate strategy pursued so far. Determination of necessary capabilities and other important factors in SWOT and VRIO analyses, which are tools to assess a company's strengths, weaknesses, opportunities, and threats, as well as the qualities it possesses that are valuable, rare, costly to imitate, and actively exploited by the organization.
- Strategic positioning: Selection of strategy and identification of options for the strategic position. A useful tool in this context is the strategy matrix.
- Strategy recommendation: A conclusive concentrate of the preceding exercise, which defines the ground rules for the players.

To WiMAX or Not to WiMAX

Which direction to take? The quandary is whether to WiMAX or not to WiMAX, or embrace other technology alternatives such as 3G, fiber-to-the-home, or maybe something totally radical or futuristic.

The Players

The typical traditional telecommunications service value chain has production stages that can be referred as access, transmission, services and marketing, and (some of) the new options such as QoS.

The new fields of business mainly lie within the innovative network access technologies and standards — from wireline to limited mobility networks on the basis of wireless local loop technology to 3G, wireless network, and the integrative 4G. Further extensions are possible within services and marketing, the former comprising intelligent network services and E-solutions including systems development, and the latter largely dealing with m-marketing, which again provides good topics to discuss; however, it is outside the scope of this book.

For some service providers, in developed markets declining profit margins, saturated markets, strong competition, and considerable fees paid to governments for licenses leave them with no choice other than to stride ahead toward new technologies and new sources of income. Although it may seem promising, this path is anything but easy.

For others, the new upcoming technologies are yielding many new opportunities, maybe more than they had ever sought. There is the possibility of expanding the traditional value chain in many ways, in both the horizontal and vertical directions.

Vendors see themselves in a totally different light as their business opportunities are created by hardships faced by service providers. Vendors are being driven by operators who need to increase their revenues and offset the decline in conventional telephony and Internet services, by moving into high-speed, data-optimized mobility. The second area of welcome pressure for vendors is the drive for efficiency and cost cutting by operators.

The success of any public technology initiative, including WiMAX deployment, depends heavily on a generally supportive environment, a prerequisite of which is that policy makers, the public and private sectors, and the local media have adequate awareness of benefits and other key issues relevant to the technology in question. Raising awareness and building national consensus about the benefits of low-cost broadband wireless Internet infrastructure solutions is therefore an important step.

Each country has unique characteristics and conditions with respect to BWA deployment, from a geographic, social, economic, regulatory, and telecommunications infrastructure standpoint. There are, however, several regulatory and economic factors that regulatory authorities should keep in mind while performing their duties.

As in the case of any infrastructure, WiMAX will also need substantial investments. To avoid relearning past lessons, it is vital that a wireless network be self-sustainable in some way, be justifiable, and create value or cost savings for the users. From an investment point of view, a network must be able to attract customers, maintain them, and thrive in a competitive market. Investment opportunities exist in various areas such as silicon, OEM, operators, WISPs, etc., but some areas are more lucrative then others.

Successful business strategies rely to a great extent on how well future developments are anticipated and incorporated by the chosen concept. The relevant metric is the return on investment, both in terms of the expected relative profit shares of each stage within the production process, and in terms of its absolute aggregated value. Let us examine this for access technology.

The importance of access technologies, capabilities, and capacities with wireless networks such as WiMAX will increase, not the least due to product differentiation. Although large investments will temporarily push the profit share down, in the long run, network access will play a bigger role in the value chain. Total returns from the access networks will increase owing to the rising number of customers and transactions.

Service Providers

Service providers who will be benefited immensely by WiMAX fall into three camps: wireline or fixed/mobile carriers who need to shift their business model away from fixed wires, cellular companies with no 3G license, and alternative or upstart providers who see a chance to snatch market share from the incumbent telecom and cellular companies. All this movement will force the tier-1 cellular operators, whose attitude to WiMAX is clearly more ambivalent, given their investment in 3G, to stop burying their heads in the sand and make some early decisions on WiMAX.

WiMAX is a technology that provides existing as well as new service providers a chance to widen their portfolio while increasing profitability, a value proposition unheard of in past.

New Operators

WiMAX threatens to break the incumbents' slowness in "unbundling" their copper local loop, and the lengthy discussions over suitable access charges for second operators to use these loops have been a major brake to the growth of alternative operators to the incumbent in most markets. WiMAX seem to be the "dream technology" to break that "last mile" monopoly as it has enough capacity for large numbers of customers and intensive users, a DSL-level of high-speed data access, and large coverage area, so it should allow wide-scale coverage at reasonable investments.

WiMAX allows a second operator or an ISP with such a license the choice to bring voice, data, and broadband access directly to their businesses and consumer customers presently at fixed locations such as the home and office, and even to the access base station of a W-LAN public hot spot without needing the cooperation of the incumbent carrier. By eliminating access charges, the incumbent will be under pressure to do tariff rebalancing. So it would change the balance in the market.

The second operator will be both free from the technological constraints of copper local loop and the installation times of the incumbent, as well as in control of how much capacity to install where and when, and which mix of services to offer to which customer. WiMAX therefore promises to vitalize the competition between incumbents and second operators.

Longer-term extension of WiMAX to voice services and then to mobile services would strengthen this competitiveness even further, and broaden it to make WiMAX an alternative to mobile voice and 3G data services. There is little remaining need for regulation of telecoms, as with "fair" competition throughout the network, normal consumer protection laws should be adequate.

Incumbents

Mobility vastly increases the potential revenue and profit margins to be derived from fixed wireless, opening up customer bases on a national scale, supporting new applications such as interactive gaming at high speed, and giving WiMAX operators the chance of providing a service that is differentiated from DSL and cable even in urban areas. But all classes of operators are only just beginning to study the economics of investing in WiMAX, which will depend on the cost of CPE as well as infrastructure cost. The other key consideration on the infrastructure front is where to mount the base stations. This is where cellular operators have an advantage over their wireline cousins, because they already have a network of towers and other real estate on which to position WiMAX access points. For companies without a 3G license or those that are daunted by the cost of rolling out UMTS or EV-DO, the argument for deploying WiMAX as a parallel, high-speed network on existing infrastructure can be compelling. The biggest example of this thinking is Nextel in the United States, with its broadband wireless trials aiming to supplement its ageing iDen cellular network, but there will also be strong opportunities for regional and tier-2 cellular companies in many areas of the world.

The advantages are enhanced still further for cellular companies when they can access base stations that integrate broadband wireless, cellular networks, and backhaul for both technologies in one structure. This will shift the economics of running two parallel networks for tier-2 cellular companies and potentially outweigh the tier-1 carriers' fears of hurting 3G revenues with such a system. The importance of such integration is illustrated by the entry into the WiMAX community of companies specializing in multi-network or software-defined base stations.

Equipment Vendors

First, the vendors must try to get their products into the market at an early stage, and try to prevent non-WiMAX BWA solutions from gaining too strong a foothold to dislodge. Such non-WiMAX solutions are being driven by operators that need to increase their revenues and offset the decline in conventional telephony and Internet services, by moving into high-speed, data-optimized mobility. This existing demand is a powerful attractive factor for the makers of broadband wireless equipment.

Second, until recently, mobility was the preserve of a few specialist vendors, many of whom have now shifted into the WiMAX camp and are promising migration paths to 802.16e — Navini and NextNet being good examples. All long-term WiMAX players must take up mobility very seriously as it has the potential to be the biggest opportunity of their lifetime.

Third, another promising area is the CDMA and GSM networks' upgrade paths to the future mobile 802.16e specification. One of the vendors recently acquired by Alvarion, interWave, offers a similar solution, "network-in-a-box," which collapses the functionality of all three — a softswitch, base station controller, and mobile switching center elements — into a single rack for GSM applications. This provides a turnkey solution for carriers, and will be valuable in markets in which budgets are tight, especially rural or developing regions where populations may be sparse and carriers small.

Regulators

BWA technologies and their applications are in their early stages of development, and their potential economic and social benefits appear to be considerable. These new technologies and applications, however, are emerging in many different situations, often outside the operational landscape of traditional telecommunications services, and with new types of participants from both the private and public sectors.

BWA initiatives may be seen as disruptive and can hit unintended roadblocks in the form of local regulations and lack of understanding of their potential. Conversely, proper government support and incentives can accelerate their successful implementation at little cost and with significant immediate economic and social benefits for the poor.

Each country presents unique characteristics and conditions with respect to BWA deployment, from a geographic, social, economic, regulatory, and telecommunications infrastructure standpoint. There are, however, several regulatory and economic factors that regulatory authorities should keep in mind while performing their duties. We will discuss some of the key issues that have a direct bearing on the success of BWA initiatives as well as the resultant sustained balanced economic development.

Encouraging the Aggregation of Demand for Bandwidth

One of the most important factors of success for wireless ISPs is a rapid increase of initial demand for connectivity, which allows for a faster breakeven on operating expenses. This can be achieved through initial aggregation of demand based on applications for local public services such as schools, universities, health services, and public administration. Business, agriculture, and private uses will inevitably add to the mix once service is available. In many underserved areas, it is likely that initial viable aggregation will occur through the deployment of wireless Internet kiosks operated by small entrepreneurs.

Identify, Promote, and Establish National Consensus

The success of any public technology initiative, including BWA deployment, depends heavily on a generally supportive environment, a prerequisite of which is that policymakers, the public and private sectors, and the local media have adequate awareness about benefits and other key issues relevant to the technology in question. Raising awareness and building a national consensus about the benefits of low-cost broadband wireless Internet infrastructure solutions is therefore an important step.

Identifying leading applications, which may drive the initial use of wireless Internet infrastructure and distribution, will further develop support for wireless Internet solutions among key constituents. Governments should also encourage local public services to use the infrastructure of local wireless ISPs. Local governments may also inventory those applications that may contribute the most to bridging the digital divide, both from a geographic and social standpoint, and foster economic development, job creation, and productivity gains in all economic sectors.

Create Environment for Collaboration

It is anticipated that wireless technologies and applications will continue to evolve rapidly, which makes it important for governments and private-sector leaders to remain abreast of other countries' experiences, regulatory work at the international level, best practices, and latest innovations. Governments must encourage knowledge sharing among their own constituents as well as with other countries to leverage the existing knowledge and experiences of other organizations. Sharing of ideas can be very productive, especially in the areas of E-government, E-education, and E-health.

Wireless Internet infrastructure and services can leverage a number of existing resources in any given country. Some of the areas and players that can be considered for fostering cooperation are the following:

- Backbone (operators and owners of fiber-optic networks including governments, private-sector networks, telecommunications companies, power-grid operators, and satellite communications operators)
- Location (owners of real estate and high points with adequate power supply and security to install antennas, such as existing radio communications towers or public sector buildings such as post offices or other types of standard venues)
- Expertise (systems integrators with the technical capabilities to install and maintain wireless equipment such as towers, cabling, hub-integrated antennas, wireless modems, control and network management systems, routers, cables, uninterruptible power supplies, racks, etc.)
- Know-how (operators of similar services such as TV broadcasting, cellular telephony, computer maintenance organizations, and power distribution)
- Funds (finance companies interested in funding start-up ISPs)

Such initiatives can lead to shortening of project cycles as well as decreasing cost, improving the chances of success.

Investors

After a long time investors are finding some solace as well as appeal in the telecommunications sector. Having burned their fingers in the telecom bubble bust, caution is bound to be the buzzword. One strong point in favor of WiMAX is that wireless economics have already been proved true in mobile voice services, in which the cost per line is 40 percent of the equivalent cost for wireline services. The same economies are expected to hold true for wireless data with WiMAX.

Investor confidence and the overall perception of WiMAX has suffered lately owing to the delay of about six months in the 802.16-2004 certification process. Similar to an overexcited child, the technology industry is never able to stop itself wishing away the days until Christmas. Time and again, standards and products fail to meet their schedules, not because of any intrinsic problem, but because the deadlines were unrealistic to begin with.

The delay is a blow that is more symbolic than real. However, it will prompt skepticism about the program, and this will be far more serious if it has a knock-on effect on the upcoming mobile standard, 802.16e. It is essential that from now on the WiMAX Forum set realistic deadlines and not allow any further risk of missing them.

Another important thing to consider is that such roadblocks, however small, can increase investors' hesitancy, which in part stems from previous broadband wireless failures such as Teligent and in part from the sheer complexity of the WiMAX business — the need for spectrum and standards, and the din of competing wireless broadband technologies such as EV-DO, HSDPA, and Wi-Fi.

Despite the uncertainties currently surrounding WiMAX deployment, even analysts who believe Intel has overhyped the technology still believe WiMAX will ultimately be successful.

Some facts that can attest to the likely future success of WiMAX or at least confirm its march in right direction is that the number of members of the WiMAX Forum has grown from less than 50 to 250 in the last year, and the number of operator members has grown from 3 to 80.

Further, similar to most other research groups, the META Group predicts that WIMAX will prove economically beneficial for alternative carriers in four key areas:

- Reducing CAPEX to under $240 per customer by 2007
- Reducing OPEX by 41 percent compared to wireline in the same period

- Reducing customer churn through improved customer satisfaction, especially once the truck roll is eliminated
- Allowing greater differentiation of services

Mobility: The Catch-22

Another catch-22 situation for both operators and vendors is mobility. The big decision is whether to support the current fixed standard, 802.16-2004, at all, or to move directly to 802.16e. Although there are clear short-term reasons to invest in fixed WiMAX in traditional broadband wireless markets such as T1, microwave replacement, or rural access, the large vendors that operate in the cellular world are tending to wait for the "e" standard before making large commitments. Some heavyweights treading this path are Motorola and Alcatel, among many others.

Until recently, mobility was the preserve of a few specialists, many of whom have now shifted to the WiMAX camp and are promising migration paths to 802.16e — Navini and NextNet being good examples. Now most of the vendors are inclined toward mobility. Many large telecommunications companies today see WiMAX as an opportunity to add the option of mobility as they move toward converged networks. These giants will want mobility or at least portability from vendors.

Mobility vastly increases the potential revenue and profit margins to be derived from fixed wireless, opening up customer bases on a national scale, supporting new applications such as interactive gaming at high speed, and giving WiMAX operators the chance to provide a service that is differentiated from DSL and cable even in urban areas.

The classes of operators are beginning to study the economics of investing in WiMAX, and for them, cost will be the most important factor in making decisions. The question then is how quickly and how low subscriber unit prices and the cost of infrastructure will fall.

Getting the Best out of WiMAX

As with the introduction of any new disruptive technology, it is important for operators and carriers to understand that there are a number of variables factored into their WiMAX deployment strategy.

Operators and carriers who wish to get a leg up on the competition by choosing to deploy WiMAX-ready platforms today should be able to minimize their risks and maximize their market opportunities by keeping the factors described in the following subsections in mind.

Planning

A WiMAX network privately owned by a commercial operator will be made available to consumers, businesses, and municipal, state, and federal institution and agencies as well as other entities.

Existing Environment Assessment

The first step in planning any network from a business perspective is to inventory the existing environment, including the following factors:

- A thorough analysis of current telecom offerings, both wired and wireless, including price points, service offerings, and value-added benefits to the customer
- An inventory of current and future frequency options
- An inventory of existing assets including building rooftops, city-owned towers and poles, existing telecom use, and other ways to improve existing services

Making Assumptions

After assessing the existing environment, the second step would be to make some broad assumptions about the potential use of the network. This effort will assist in answering basic questions such as the estimated bandwidth needed as well as the number of cell sites needed. Applications can include the following:

- Fixed best efforts or carrier-class data connectivity
- Fixed Voice-over-IP via wireless services
- Mobile data services
- Backhaul to public and private hot spots

Plotting Projections

This involves determining the number of potential customers in the market, the anticipated five-year penetration rate (given the current level of competition in the market), and the potential customer churn rate. Based on the planned service offerings (discussed earlier), an operator can estimate retail pricing.

Once the potential customers, their uses of the network, the prices charged, and the estimated cost of the network are identified, remaining details such as Internet backhaul, billing costs, technical support costs, and maintenance and monitoring costs can be estimated.

Presenting Business Case

Finally, returns and costs can be compared to understand profitability as well as break-even points or even returns on investment, internal rate of return (IRR), NPV, months to cash flow positive, and other financial metrics.

Deployment

A low-risk WiMAX-ready solution must take into account a number of factors which include choosing a product or technology platform that has the following attributes:

- It is field-proven: If a carrier decides to deploy WiMAX equipment today, it must seek out technology — and an equipment provider — that has already demonstrated its robustness, stability, capabilities, and efficiency over time in real-world situations. Testing a new air link is a simple affair, but scalability must be proved in the field. Moreover, the importance of meticulous RF planning, network architecture, and network management cannot be overstated, and this requires choosing a vendor or system integrator that has also been field-proven over time.
- It is cost-effective: Carriers cannot take the risk of deploying a product or solution that simply promises to reduce its cost sometime in the future. Costs must be in line with the carrier's immediate business plan.
- It provides a flexible, long-term upgrade path: WiMAX technology has the capability to enable networks on the scale of today's mobility networks, with thousands of base stations serving millions of subscribers. Successful vendors will offer a flexible, cost-effective migration plan for such networks that is flexible enough to adapt to evolving WiMAX-standard profiles. In other words, insofar as possible the WiMAX-ready platform should be future-proof.

It meets both present and future needs: Carriers should seek out a BWA solution that can propel their service offerings both today and tomorrow. This requires a thorough analysis of one's specific business case, a careful evaluation of the WiMAX-ready products that are currently on the market, and an assessment of the WiMAX profiles that currently exist to determine which one best addresses the realities of the carrier's evolving markets and business plan.

Recommendations

Service Providers

Opportunities abound for service providers to bring custom wireless Internet solutions to vertical markets such as healthcare, hospitality, utilities, real estate, retail, warehousing, field service and sales, and last mile communications. In addition to these vertical markets and home and office networking, service providers also have excellent opportunities to operate public (free) and private (commercial) networks, or hot spots.

For new operators, WiMAX is a dream technology and they should lap it up.

Incumbents must assess existing infrastructure as well as services offered, based on what they must deploy WiMAX to enhance portfolio as well as reach of the service.

Equipment Vendors

Vendors will see WiMAX and the mesh network product market as providing excellent opportunities. More multimillion dollar networking projects will be seen in the next few years than ever before. The major thrust would come from metropolitan markets, especially those run by municipalities.

On the operator side, more new players will emerge, with most of the buying over the next few years coming from players not even existing in the telecom operations landscape today, and the other big segment would be non-3G-cellular companies.

As for the type of equipment, the inclination would likely be heavily toward the unlicensed and mobile segments. The need for integrated infrastructure, i.e., mobile (GSM/CDMA) and broadband will rise steadily.

The key to success will be speed in delivering new products. One of the key methods of accelerating uptake of a new technology is the development of reference platforms, which reduces time to market for OEMs. Depending on the area of expertise, vendors must come out with such solutions at the earliest.

Regulators

Regulators should update telecommunications regulations to foster market opportunities, optimize existing infrastructure resources, and free competition among wireless service providers. In our dynamic world, situations vary greatly from one country to another, still keeping in mind that there are four activities that need careful attention from a telecommunications regulatory perspective:

Identify and facilitate availability of key existing resources and foster cooperation among potential actors.

Assist in skill and capacity development by initiating focused learning programs.

Adopt minimum regulations supporting the use of unlicensed spectrum and ICT industry standards.

Support the experimentation of new services.

What Works

The following would have a beneficial impact on the BWA market:

- Government policy initiatives to maximize ability of all citizens to use BWA
- High-level commitment to policies aimed at promoting competition as competition drives higher speeds and lower tariff
  - Independent regulator
  - Facilities-based competition
  - Unbundling and line sharing: example being Denmark
  - Cable divestiture by incumbent telecommunications carriers: example being Switzerland
  - Ensuring spectrum is available for innovative solutions

Investors

Answering the following questions will provide sufficient guidance to finding money in WiMAX.

What technology vendors are best suited to capitalize on the market?

What is the realistic timeframe for market development?

Is now the right time to jump into WiMAX?

Should we wait for the market to develop further before trying to determine who will win or lose?

Keeping existing bigwigs in the telecommunications field dangling a little bit in WiMAX aside, there is tremendous potential for new vendors and operators embracing WiMAX in a big way. Other groups that can achieve extraordinary success are those particularly interested in WiMAX chips, applications, and roaming software. The next three years will be a fun time for everyone who is WiMAXed!

Chapter 14

Conclusion

The World Wants Access

All over the world, people want access:

- Users want access to networks.
- Network operators want access to customers.
- Users want access to diverse media simultaneously.

Broadband Access to Buildings

User concerns:

- For the last mile — fast local connection to network
- Business and residential customers:
  - Data
  - Voice
  - Video distribution
  - Real-time videoconferencing
  - Gaming applications
- Availability a must for network operators
- High cost of high-capacity cable/fiber
  - Construction costs not in line with Moore's law

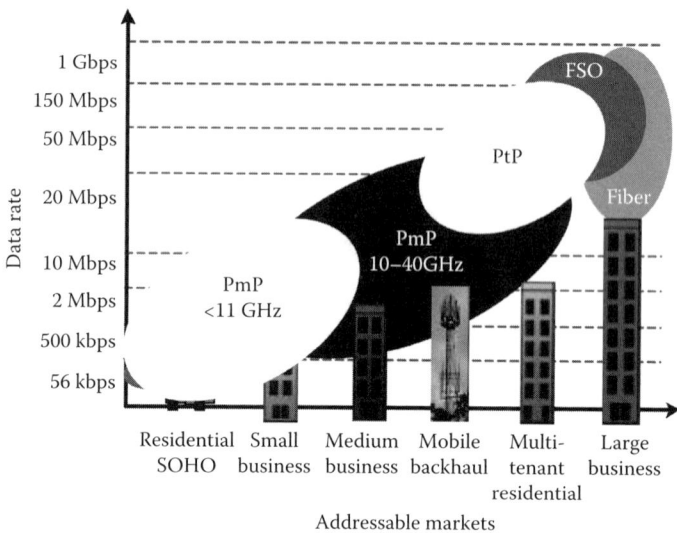

Figure 14.1 Market segments for wireless access.

Broadband Wireless Access (BWA)

BWA flourishes where:

- Many users are dissatisfied with access.
- Network operators need to reach customers.
- Distance between customers is high, whereas density very low.

WiMAX

The term WiMAX (Worldwide Interoperability for Microwave Access) has become synonymous with the IEEE 802.16 wireless metropolitan area network (MAN) air interface standard. In its original release, the 802.16 standard addressed applications in licensed bands in the 10 to 66 GHz frequency range. Subsequent amendments have extended the 802.16 air interface standard to cover non-line-of-sight (NLOS) applications in licensed and unlicensed bands in the sub-11-GHz frequency range.

Filling the gap between wireless LANs and WANs, WiMAX-compliant systems will provide a cost-effective fixed wireless alternative to conventional wireline DSL and cable in areas where those technologies

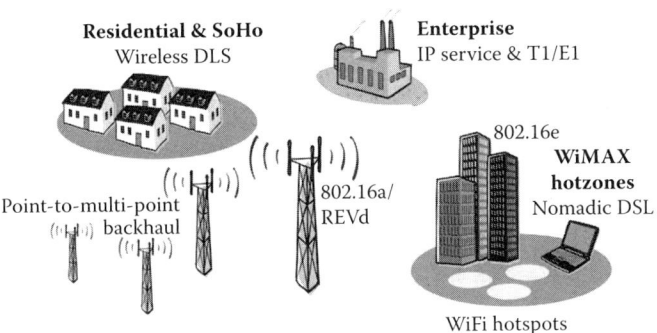

Figure 14.2   WiMAX deployments.

are readily available. And more importantly, WiMAX technology can provide cost-effective broadband access solution in areas beyond the reach of DSL and cable. The ongoing evolution of IEEE 802.16 will expand the standard to address mobile applications, thus enabling broadband access directly to WiMAX-enabled portable devices ranging from smart phones and PDAs to notebook and laptop computers (Figure 14.2).

The key features of WiMAX are summarized in the following text:

■ New standard for BWA
■ WiMAX is similar to Wi-Fi
   – Higher power increases distances
      • Miles versus hundreds of feet
■ Spectrum efficient
■ Flexible
■ Metropolitan area access technology
■ Better security, authentication, and protection against theft of service
■ Possibility of using both licensed and unlicensed frequencies
■ Quality of service
   – Triple play enabler

Properties of IEEE Standard 802.16

The following are the main features:

■ Open in development and application
■ Aimed at worldwide markets

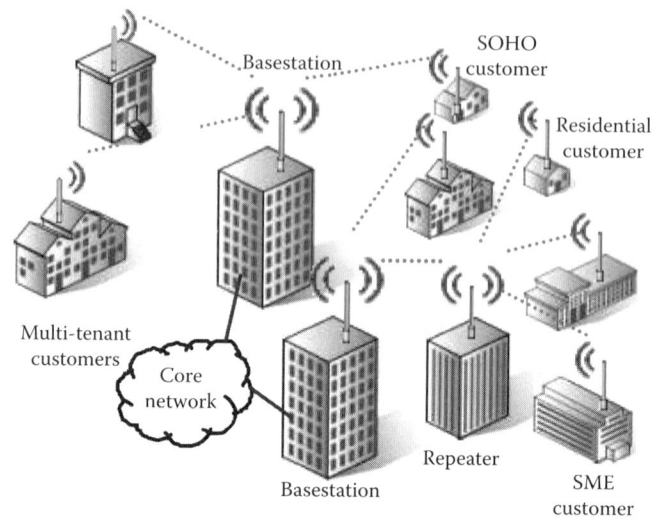

Figure 14.3   Wireless metropolitan area network.

- Engineered as optimized technical solutions
- Being enhanced for expanded opportunities
- Broad bandwidth
  - Up to 96 Mbps (>70 Mbps throughput) in 20 MHz channel (in wireless MAN TM-OFDM air interface)
- Supports multiple services simultaneously with full QoS
  - Efficiently transports IPv4, IPv6, ATM, Ethernet, etc.
- Bandwidth on demand (frame by frame)
- MAC designed for efficient usage of spectrum
- Comprehensive, modern, and extensible security
- Supports frequency allocations from <1 to 66 GHz
  - ODFM and OFDMA for NLOS applications
  - TDD and FDD
- Link adaptation: adaptive modulation and coding
  - Subscriber by subscriber, burst by burst, uplink and downlink
- Point-to-multi-point topology, with mesh extensions
- Support for adaptive antennas and space–time coding
  - Beam forming and MIMO
- Significantly complete
  - Test spec documents in development
  - Extensions to mobility coming next

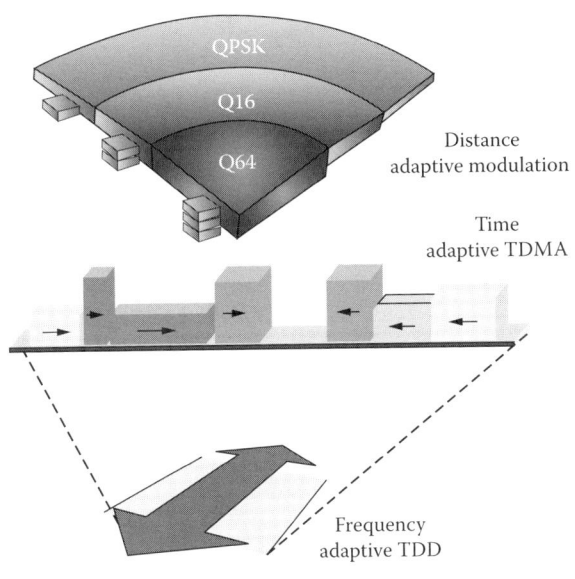

Figure 14.4   Third-generation technology in 802.16.

Third-Generation Technology in 802.16

The following are the key features of third-generation technology in 802.16 (Figure 14.4):

- Adaptive modulation: Variable modulation maximizes both air-link capacity and coverage.
- Adaptive TDMA: True bandwidth on demand and variable packet sizes provide differentiated, bursty services to multiple users.
- Adaptive TDD: Variable asymmetry in a single broadband channel best matches bandwidth to demand.
- Coverage/capacity advantage of adaptive PHY layer: Modulation changes dynamically to match propagation path conditions (Figure 14.5).
- Supports beam forming and MIMO: These provide high system gain for maximum coverage and availability and are used for extended range or increased capacity.

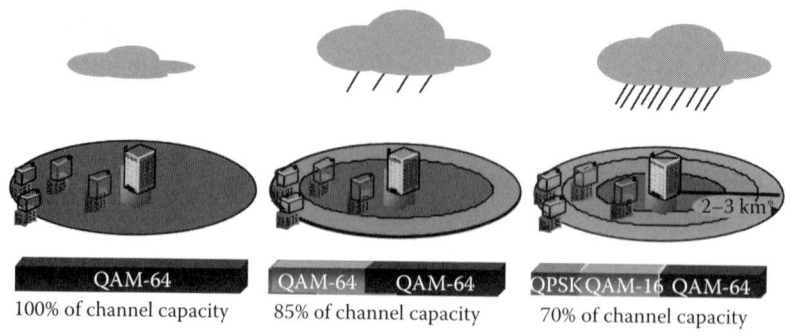

Figure 14.5   Advantage of adaptive PHY layer.

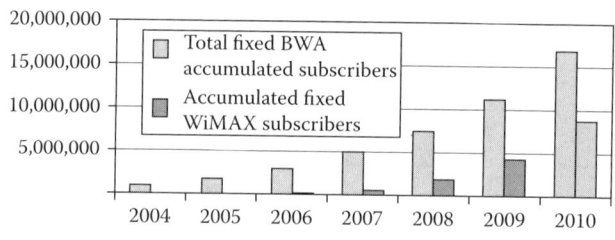

Figure 14.6   Projected subscribers of BWA.

WiMAX Forum Purpose

The following are the chief goals of the forum:

- To promote a common broadband wireless standard
- To develop reduced scope "profiles" to ease development
- To fill the gaps in the IEEE process relative to the ETSI process
- To create a BWA conformance and interoperability certification process
- To act as a certification body

Index